AF552350

STORMWATER INFILTRATION

Bruce K. Ferguson

Library of Congress Cataloging-in-Publication Data

Ferguson, Bruce K.
Stormwater infiltration/Bruce K. Ferguson.
p. cm.
Includes bibliographical references and index.
ISBN 0-87371-987-5
1. Stormwater infiltration. I. Title.
TD657.5.F47 1994
628'.212—dc20 94-11424
CIP

Direct all inquiries to CRC Press, Inc., 2000 Corporate Blvd., N.W., Boca Raton, Florida 33431.

Lewis Publishers is an imprint of CRC Press

International Standard Book Number 0-87371-987-5
Library of Congress Card Number 94-11424
Printed in the United States of America 1 2 3 4 5 6 7 8 9 0
Printed on acid-free paper

The Author

Bruce K. Ferguson is Professor of Landscape Architecture and Director of the Master of Landscape Architecture Program at the University of Georgia.

He obtained the B.A. degree from Dartmouth College in 1971. In 1975 he obtained the Master of Landscape Architecture degree from the University of Pennsylvania, where he was influenced by Ian McHarg's work in environmental planning. His particular interest in renewable resources was reflected in the subject of his thesis, the design of a self-sustaining community. He practiced landscape architecture in the Pittsburgh area from 1975 to 1980, mostly in the areas of community design and environmental assessment. He was Instructor of Landscape Architecture at Pennsylvania State University from 1980 to 1982. Since 1982 he has been at the University of Georgia, where he is a member of the Graduate Faculty, the Environmental Ethics Faculty, and the Conservation and Sustainable Development Faculty.

His specialty within the environmental area of landscape architecture has been landscape hydrology and water management since 1981. In this specialty the water balance is a unifying theme and governing principle that has led him to study the potential of stormwater infiltration.

He has written more than 100 scientific and professional publications in the areas of environment and urban development. His book, *On-Site Stormwater Management*, co-authored with Thomas N. Debo, is in its second edition. He regularly conducts stormwater management continuing education courses nationwide for practitioners of landscape architecture, engineering and planning. He has been usede as an instructor in water and environmental management by the U.S. Soil Conservation Service and the National Aeronautics and Space Administration.

He has served as a consultant and expert witness for design firms, legal firms, industrial corporations and associations, and all levels of government on landscape water conservation, urban runoff and sediment control, stream rehabilitation, wetland establishment, and mitigation of acid mine drainage.

From 1985 through 1992 he served as Secretary and President of the Council of Educators in Landscape Architecture, where he was instrumental in standardizing the council's annual *Proceedings and Landscape Journal*. In 1990 he founded the American Society of Landscape Architects' Open Committee on Water Conservation, and served as its first chairman from 1990 to 1992.

Contents

Preface

Stormwater infiltration requires new ways of perceiving the urban environment and the aspirations of stormwater management. Stormwater management is constructive only when practiced in a context of environmental interactions and the quality of urban life.

Too many American cities are caught up in stormwater management practices that are self-defeating and destructive. Too many are clinging to conventions inherited from previous generations despite growth of scientific knowledge and development of environmental values.

Stormwater infiltration is the artificial forcing of urban runoff away from surface discharge and into the underlying soil. It is qualitatively different from surface detention methods, which are treatment or delay mechanisms that ultimately discharge all runoff to streams. Infiltration causes runoff to enter a different part of the environment, there to undergo a different set of processes. Infiltration calls forth the natural powers of soil and vegetation to restore environmental health. By returning runoff to the earth, it eliminates pollutant discharge, eradicates floods, replenishes ground water supplies and restores aquatic habitats.

Infiltration obeys the same laws of physics, chemistry and biology that govern the rest of the environment. The fundamental principles of natural science provide a wealth of understanding to make stormwater infiltration accessible and feasible.

Expectations for stormwater management have evolved nationwide. Natural science has given us new realization of hydrology's interaction with urban ecosystems and new capabilities for management of the urban hydrologic cycle. Federal law now mandates control of stormwater quality: the old idea that runoff can be discharged from a property or a municipality with a mere detention basin for suppression of occasional peak flow rates, without consideration of the broader, multifaceted environment, is obsolete. Our growing understanding of the environment, the polluted quality of urban runoff, the necessity to conserve water supplies, and the failure of old stormwater systems to prevent even basic drainage problems have forced attention on alternate means of disposal and recycling. Advances in microcomputing leave us few excuses for not finishing the job of watershed and aquifer protection in the way we now know it ought to be done.

Stormwater infiltration has been practiced in numerous but widely separated locales for half a century and more. "Water spreading" to recharge alluvial aquifers has been practiced in the arid southwestern United States since the early years of this century, originally to conserve irrigation water in rural areas and still actively continued in urban watersheds such as those in the Fresno and Los Angeles areas. Long Island, New York began urban infiltration in the 1930s; monitoring by government agencies has documented beyond question its contribution to the Island's ground water supply. In the mid-1960s, runoff was infiltrated into individual residential lots in El Paso. In the 1970s, the Minnesota Department of Transportation developed the water balance as an analytical approach to infiltration design. In the 1980s, the state of Maryland created an environmentally ambitious statewide infiltration program. For decades Florida has been practicing infiltration to protect water quality in a sensitive geohydrologic setting.

Today there are more than ten thousand infiltration basins in the United States. Infiltration has been practiced in deserts, Mediterranean climates, humid tropics, temperate zones, and the frigid North. Runoff has been infiltrated into glacial till, dune sand, alluvial plains, karst sinkholes, and the residuum of sedimentary, metamorphic and igneous rocks. It has been practiced in forests, plains, deserts, chaparral, savannah, and scrub environments. It has been combined with land uses of single- and multiple-family residences, industries, shopping centers, golf courses, ball fields and artificial wetlands. Of 77 departments of transportation and public works surveyed throughout the United States, 26 reported use of infiltration (Hannon, 1980, Appendix A). It has been practiced alone and in combination with stormwater conveyance, detention and wetland systems. Its implementation has prevented or solved problems of flooding, storm sewer cost, water quality, wetland maintenance, and replenishment of ground water and stream base flows. In arid regions it has been combined with water harvesting to reduce irrigation demands of urban plantings.

Infiltration's status is founded upon a body of scientific knowledge accumulated during decades of hydrologic modeling, laboratory experimentation and field experience (Ferguson, 1984 and 1989). Technical reports have been produced by United States federal agencies including the Agricultural Research Service, the Environmental Protection Agency, the Federal Highway Administration, the Geological Survey and the Soil Conservation Service. Studies have been sponsored by water resources research institutes in the states of Florida, Georgia, Hawaii, Maryland, North Carolina and Texas. Results have been published in the *Journal of Environmental Management*, the *Journal of Soil and Water Conservation*, *Landscape Architecture*, the *Proceedings of the American Society of Civil Engineers*, the *Transactions of the American Society of Agricultural Engineers*, and *Water Resources Research.* Design recommendations have been made by highway departments of the states of California, Florida, Minnesota and New York, the Department of Water Resources of the state of Maryland, the Washington State Department of Ecology, and construction associations including the American Iron and Steel Institute, the American Public Works Association, the Florida Concrete and Pavement Association, and the Georgia Crushed Stone Association.

Unlike any other approach to stormwater management, infiltration is capable of solving all the problems of urban runoff: peak flow, base flow, streambank erosion, ground water recharge and water quality. Infiltration is not just a means of mitigating the hazardous aspects of stormwater; it is a means of reclaiming water resources and rehabilitating urban watersheds. Infiltration deserves to be the beginning of stormwater management, the fundamental tool for solving stormwater problems, the approach that is implemented to its fullest feasible extent before anything else is attempted.

The keys to successful infiltration design are the same as those for any successful environmental design, including basic training in one's technical, environmental or design discipline, active awareness of the broad range of topographic, geological, hydrological, biological, social, aesthetic, construction, and maintenance influences that play on any site-specific project, and ingenuity and care to develop an economical system that adapts to them.

Despite infiltration's advantages, it has not been practiced in many parts of the United States. The biggest impediment to its use has been mere lack of awareness. In any human endeavor, of all possible reasons for failure, the worst is ignorance. Those involved with land management, site and drainage design, urban development and environmental management ought to possess infiltration as one of the tools they can employ daily if they wish to be part of the ecological solution and not part of the problem.

This book's purpose is to open the window to infiltration so that everyone can see how brightly the light shines upon this waiting jewel. It presents the scientific concepts that make infiltration important and feasible and illustrates the wealth of experience that undergirds this gentle technology, while frankly pointing out the problems that have occurred in the field so that the mistakes of the past need not be repeated.

The book is not a basic introduction to natural science or stormwater management. It is for students and practitioners in the design professions and environmental management, who are beyond the introductory levels of their disciplines and who now perceive a need to go farther into this specific and—for many—new area. Readers needing background in general stormwater hydrology and management are referred to Ferguson and Debo's (1990) basic textbook.

This book places stormwater management in a holistic context. Its material is drawn from a number of disciplines such as agronomy, forestry, geography and geology. It is necessary to overcome disciplinary boundaries in order to understand science and conduct design at their best.

The author has been studying stormwater infiltration and its technical, environmental and land use contexts for 20 years. Along the way, Narendra Juneja, Malcolm Wells, Michael Dufalla, Richard Parizek, John Hewlett, Robert Carver and Wade Nutter have contributed to the development of the concept of infiltration, or criticized the concept constructively. The Georgia Crushed Stone Association (George Howard, Executive Director) supported research. J.E. Ayars (U.S. Agricultural Research Service) and Donald F. Nichols (Los Angeles County Flood Control District) shared information freely. Two students, Stacy Patton and Morgan Ellington, contributed strongly to refinement of ideas through their dedicated and conscientious research.

For this book, Deborah Dalton, Thomas N. Debo, Robert L. Kort, Forster Ndubisi, David B. Nichols, Charles A. Pryor, Todd C. Rasmussen, Earl Shaver and Malcolm E. Sumner generously helped reduce technical and stylistic errors by reviewing one or more draft chapters. Tamas Deak digitally scanned and printed photographs for publication. Kris Larson checked a draft manuscript for grammar and clarity. Roger D. Moore generously provided Figure 2.10 and permitted its publication.

Urban communities, once built, tend to stay in place, occupied and functioning in one way or another, for many human generations. Decade after decade, the rain falls, the trees grow, and the decisions of form and process made by designers remain. In the construction of cities there is time for reflection, for meticulous searching out of unexpected alternatives, and for revision of decisions previously made. The entire design and environmental community must move the state of the practice closer to the state of the science. The goal is to solve environmental problems, by whatever method is supported by the evidence. In every project we must try to restore biophysical integrity, to reintroduce the mechanisms of natural equilibrium. We must seek new and better ways with an open mind, a humble heart, and a will of iron.

REFERENCES

Ferguson, Bruce K., 1984, *Infiltration and Recharge of Stormwater: A Resource Conserving Alternative for the Urban Infrastructure*, P-1563, Monticello, Illinois: Vance Bibliographies.

Ferguson, Bruce K., 1989, *Urban Stormwater Management Bibliography*, P-2795, Monticello, Illinois: Vance Bibliographies.

Ferguson, Bruce K., and Thomas N. Debo, 1990, *On-Site Stormwater Management, Applications for Landscape and Engineering*, second edition, New York: Van Nostrand Reinhold.

Hannon, Joseph B., 1980, *Underground Disposal of Stormwater Runoff, Design Guidelines Manual*, FHWA-TS-80-218, Washington, D.C.: U.S. Federal Highway Administration.

1

Subsurface Water in the Hydrologic Balance

The unique importance of stormwater infiltration comes from the role of subsurface water in the hydrologic balance. The defining process of infiltration is to transfer surface water to the subsurface. Water occurs in the subsurface of almost all landscapes. All subsurface water has an intrinsic sustaining role in landscapes and watersheds. Understanding how water moves through the subsurface provides a foundation for evaluating the impacts of urban development on landscape hydrology, and planning stormwater management in response to them.

In 1580 Bernard Palissy overthrew the medieval myth that rivers flowed magically from the center of the earth by concluding that rivers "take their origin in and are fed by rain alone." He explained (Adams, 1938, p. 448), "...rain waters rushing down the mountain sides pass over earthy slopes and fissured rocks and sinking into the earth's crust, follow a downward course until they reach some solid and impervious rock surface over which they flow until they meet some opening in the surface of the earth, and through this they issue as springs, brooks or rivers, according to the size of the exit, and the volume of water to be discharged." Palissy's idea was confirmed quantitatively late in the seventeenth century, when Claude Perrault measured the flow of the upper Seine and found it to be only one sixth of the precipitation computed to fall on the basin, and correctly surmised that losses caused by evaporation might account for the difference.

Today we know that, on the average for the continental areas of the earth, there are 72 cm/yr of precipitation, which is partitioned into 41 cm/yr of evaporation and 31 cm/yr of runoff (van der Leeden, Troise and Todd, 1990). Below the ground surface, water soaks into the void spaces of porous material, and flows from pore to pore, around the surfaces of the encompassing solid particles. Subsurface water can be divided generally into that in the unsaturated material immediately below the ground surface and that in the underlying saturated material. The boundary between the two zones is the water table.

ELEMENTS OF SUBSURFACE HYDROLOGY

Earth materials

Every earth material contains some amount of pore space—the primary pores or micropores—inherent in the type of material. Sedimentary rocks such as limestone and sandstone contain pores between the grains of cemented material. Crystalline rocks such as granite and gneiss contain minute pores between different crystals. Surface soils contain pores between the grains of sand, silt, clay and organic matter.

Many additional openings, the secondary pores or macropores, develop in response to outside forces. Open faults and joints can vastly multiply the pore space of consolidated rock. Water moving through the cracks enlarges initial openings by dissolving and eroding the surrounding material. In soils, colloidal materials (clay particles, humic substances, and inorganic colloids such as oxides of iron and aluminum) cement primary soil particles into larger aggregates, with larger, more connected pore spaces. Great volumes of connected pore space can be opened in soils by the extension of plant roots, the burrowing of fauna and the cracking of drying mud.

Table 1.1 Porosities (percent by volume) in typical earth materials (data from Beven and Germann, 1982; Brady, 1974; and Heath, 1983).

	Porosity		
Material	Primary (micro)	Secondary (macro)	Total
Clay	50	Variable	50+
Loam	40	Variable	40+
Sand	25	Variable	25+
Mixed gravel	20	0	20
Limestone	10	10	20
Sandstone	10	1	11
Basalt	10	1	11
Granite	0	0.1	0.1

A material's porosity is the volume of pore space as a proportion of total volume. Total porosity is the sum of primary and secondary porosities. Table 1.1 lists the porosities of some typical earth materials.

The void spaces in earth materials are usually filled partly with water and partly with air. When water infiltrates into the soil from rainfall, it displaces some of the air and fills a greater proportion of the pore space.

Surprisingly different from a material's porosity is its permeability, the rate at which water can be made to move through it. A material of high porosity can have low permeability if its pores are little connected to each other or if its individual pores are so small that water encounters great friction in passing through.

Pore space is generally greatest near the earth's surface and less abundant at greater depths. At some depth, usually a few hundred or a few thousand meters under the surface, cementation and the pressure of overlying materials effectively close any openings that may once have been present; then openings are rare, and the movement of liquid water becomes essentially impossible. Water seeping down from a rain-soaked surface collects above the impermeable layer, filling all the pores of the permeable portion until it overflows into the streams and oceans.

Forces acting on subsurface water

The forces acting upon water in porous earth material include the earth's gravity, the attraction of water molecules to each other and to surrounding solid surfaces, the weight of overlying water, the attractive force of ions dissolved in water, and the weight of any overlying unconstrained soil particles (Brady, 1974, p. 168-172; Jury, Gardner and Gardner, 1991, p. 48-52). In each of these force fields the water possesses potential energy, measured by the force required to move the water directly against the force. The total difference of potential energy from point to point in the earth—the sum of the various potentials or heads—determines the direction and rate of water flow.

The weight of overlying water is the pressure head p. It exists where the earth material is saturated and represents the positive pressure of the saturated water mass. It can be referred to as piezometric potential, from the Greek piezein, to press.

The attraction of water molecules to each other and to solid surfaces is the matric suction ψ (psi), in reference to the matrix of pores and water in a soil mass. It has sometimes been referred to as capillary potential. It exists where the pores are less than saturated with water. Matric suction represents the negative pressure associated with the attraction of solid surfaces for water and of water molecules for each other. It is conventionally assigned positive value for negative pressure, $\psi = -p$. Thus a larger suction is a more negative pressure.

Because the matric and pressure potentials represent a single gradient from negative through positive (unsaturated to saturated conditions), their sum can be thought of as a single pressure term. This sum is positive in saturated conditions and negative in unsaturated conditions.

The force of gravity is the gravitational potential z. It may be positive or negative, depending on the elevation chosen as a reference. If the water in question is below the reference elevation, work must be done to raise it, and the gravitational potential is positive; if above, gravity potential is negative.

The attractive force of ions dissolved in water is the osmotic potential. It results from the hydration of ions in water and can be important where subsurface water is highly saline or mineralized. Water molecules are attracted to ions in the solution. Water is pulled from a zone of less ion concentration to one of greater concentration.

The weight of overlying unconstrained soil particles is the overburden potential. It exists where soil is free to move, as in soils with high shrink-swell activity.

If osmotic and overburden potentials are assumed negligible, total potential or head H is equal to the sum of matric, pressure and gravity potentials. In saturated conditions, $H = z + p$. In unsaturated conditions, $H = z - \psi$.

The vadose zone

Immediately below the ground surface, outside of wetland areas, the earth pores are unsaturated. This part of the subsurface is called the vadose zone, from the Latin vadosus for shallow, akin to wade as in a shallow ford. Here matric suction is active: water is in tension. Water molecules are in a sense suspended among the pores of soil and rock by their attraction to solid surfaces and to each other. The amount of water suspended in the vadose zone can be vast, amounting to 48 percent of the world's non-frozen fresh water (Schultz and Hewlett, 1978).

Suction increases as water content declines. During desorption the large pores empty first; water is held longest in the small pores and on particle surfaces.

Conventional wayposts in moisture level during desorption are identified in Figure 1.1 (Brady, 1974, p. 192-199; Hewlett, 1982, p. 52-55). For the first two or three days after a thorough soaking, water occupies the larger soil pores and moves readily under the force of gravity. This gravitational water drains rapidly down through the soil with little restraint from matric forces. When rapid drainage has ceased, the soil's moisture content is considered to be at field capacity. Field capacity is conventionally assumed to be at suction of 1/3 bar. Below field capacity, water is held only in smaller pores. It continues to drain slowly from the soil, approaching an equilibrium moisture content. The permanent wilting point is the soil moisture content below which plants can no longer maintain turgid leaves; they may then go dormant, defoliate or die. The permanent wilting point of many plants is at suction of 15 bars. The wilting point is lower for plants adapted to dry conditions—as low as 75 bars for the desert shrub mesquite. Below the permanent wilting point water forms thin films, perhaps only a few molecules thick, around soil particles. It is bound so tightly that it es-

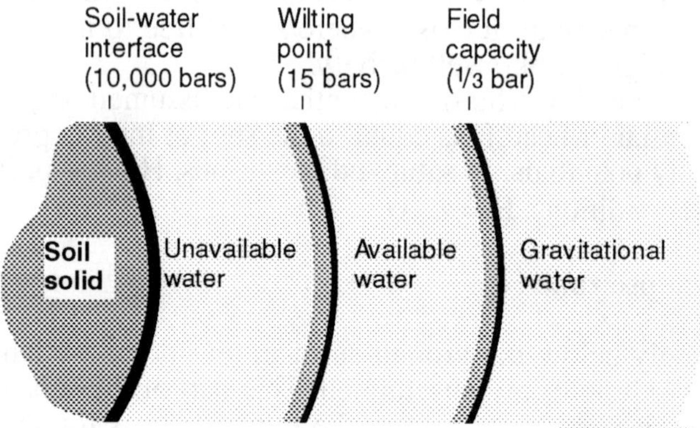

Figure 1.1 Variable matric suction of the moisture film at the surface of a soil solid.

sentially cannot flow as a liquid. Higher plants cannot absorb this water, although some microbial activity has been found to take place there.

Figure 1.2 shows the relationship of matric suction to water content during desorption of four typical soils (each curve would have a slightly different shape during wetting). A sandy soil drains rapidly after a heavy rainfall, leaving only a small amount of water clinging as a film around soil particles. A finer-textured soil has many times more surface area in its void space than a sandy soil, so it can hold more water at a given matric potential.

Available moisture is the water held between the field capacity and the permanent wilting point. It is thus an index of water available to plant life and evapotranspiration. Generally silt and clay loams have the greatest capacity to hold water in the available range, coarse sand has the least, and pure clay is intermediate.

The matric suction in a given soil mass varies from time to time depending on wetting and drying cycles, and from place to place within the soil depending on the directions in which wetting and drying have been proceeding. Soil moisture moves downward, upward or laterally in response to relative total potential gradient.

The phreatic zone

At most places on the earth, beginning at some depth and continuing down to the impermeable base of all significant water movement, the earth pores are effectively saturated with water. The saturated zone is referred to as the phreatic zone, from the Greek phreatos, meaning well. Water in this zone is commonly called ground water. The amount of ground water can be vast; worldwide it amounts to 48 percent of all non-frozen fresh water (Schultz and Hewlett, 1978). The boundary between the vadose and phreatic zones is the water table.

From time to time drainage through the vadose zone adds water to, or recharges, the immediately underlying ground water. Recharge partly or wholly replaces water lost to discharge, and at least temporarily raises the water table.

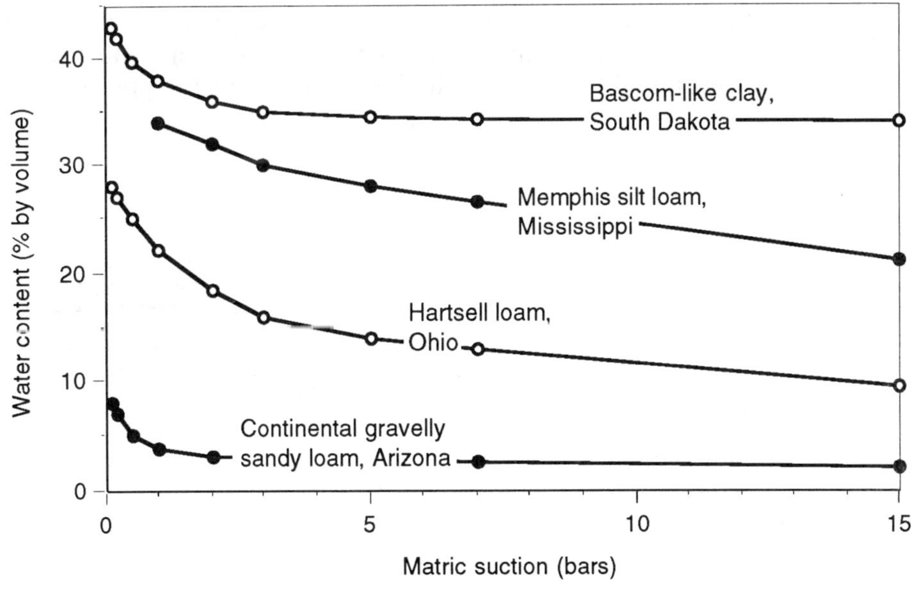

Figure 1.2 Variation of matric suction with water content during desorption of four soils (after Baver, Gardner and Gardner, 1972, Figure 8-3).

A cove or swale on a hillside, a stream channel in a valley, or any low or concave topography can incise into the water table, providing a point at which ground water discharges to a surface stream. Overall the water table slopes downward from the interior of a hill to each discharge point. Streams can potentially carry water away thousands of times faster than ground water can move through the constricted pores of earth material, so the water table at the discharge point does not rise above the swale or stream where water drains out, despite fluctuations elsewhere in the water table.

In the phreatic zone the pressure potential is positive. The greater the depth below the water table, the greater the pressure. If a horizontal plane is cut through a phreatic zone, the total head varies from place to place in the plane because the undulating water table causes different depths of water over the plane. Figure 1.3 shows two points below a hill, having two different depths and thus different heads. The slope or gradient of the water table indicates the rate of change of depth, and thus of head, from place to place. Ground water obeys the head by flowing in the direction of the steepest gradient. The mass of ground water tends to slump from the higher head toward the lower, and thus toward the low-lying discharge point.

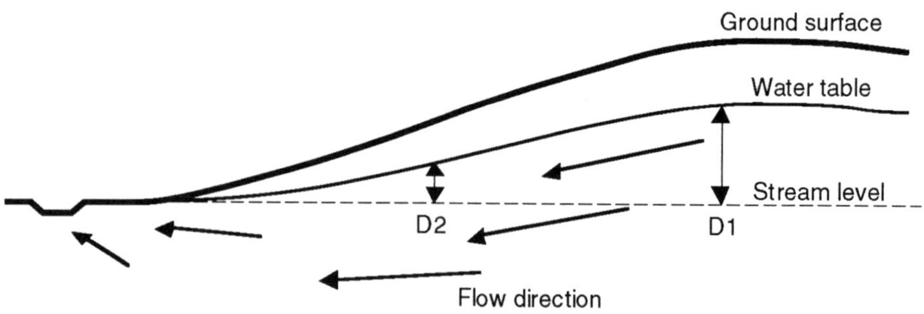

Figure 1.3 Ground water flow toward a discharge point resulting from the gradient of total head. D1 and D2 are different depths of water, and thus different total heads, creating the gradient.

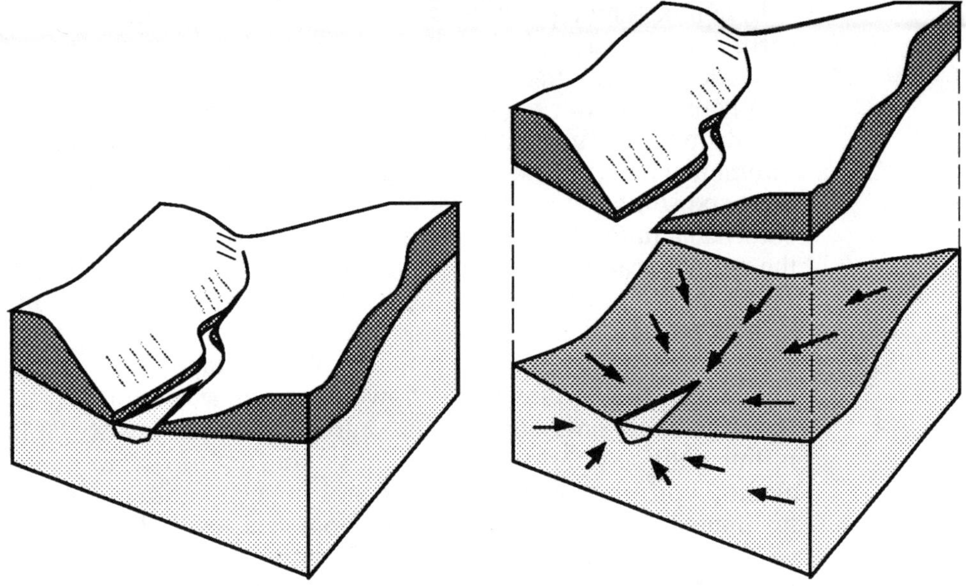

Figure 1.4 Relation of ground surface to water table (after Leopold, 1974, page 21).

The phreatic zone is continuous over large areas of watersheds. Where the earth material is homogeneous, the ground water forms a mound faintly reflecting the surface topography, with its low point at the surface discharge and its high point in the interior of the hill more or less below the topographic ridge line. Figure 1.4 shows the relation of the water table in a homogeneous earth material to the land surface where a stream channel has water flowing into it because the channel intersects the water table. The right-hand part of the figure shows the vadose zone lifted up to exhibit the undulating surface of the water table faintly reiterating the ground surface.

Units of earth material that contain water in relatively great quantity and transmit it readily, particularly those sufficient to be significant sources of water for human use, are referred to as aquifers, from the Latin aqua for water and -fer from a word meaning to carry or to bear. Examples are strata of gravel or sand, layers of sandstone or limestone, bodies of alluvium in stream valleys, and zones of fractures in otherwise slowly permeable rocks.

In contrast, an aquiclude is a zone of relatively impermeable earth material such as clay, shale or granite. An aquiclude may separate two aquifers.

Where an aquifer passes below an aquiclude, it is referred to as a deep or confined aquifer. Confined aquifers are common in sedimentary regions such as large parts of the American midwest and southeastern coastal plain. The recharge area, where a confined aquifer outcrops and can receive inflow of water from the surface, can be distinguished from the remainder of the aquifer below other geologic units. The water table in the recharge area is like water tables anywhere else, discharging at springs and faintly reflecting the local topography. As the remainder of the aquifer dips below an aquiclude and out of the recharge zone, pressure potential builds up, but the water table in this confined part of the aquifer is not free to rise to an elevation corresponding to its pressure. In such aquifers we can distinguish a piezometric or potentiometric level (from Latin words for pressure and power) from the water level in the aquifer. The piezometric level is the elevation water rises to in a well; in an aquifer under pressure this level is higher than the top of the aquifer. The greater the elevation of the recharge area relative to the part of the aquifer where the well penetrates, the greater the pressure. If the water is under such pressure that it rises in a well to the ground surface, it is called artesian, from the French province of Artois where this phenomenon was found in the days of the Romans. If the piezometric elevations in a number of wells are connected with contours, they form a piezometric surface showing the variation of pressure from place to place in the aquifer. As anywhere else in the phreatic zone, water contained in such an aquifer moves in the direction representing the steepest potential gradient.

THE HYDROLOGIC BALANCE

Subsurface fluxes

When rainwater reaches the ground or when surface water flows through a rill or is ponded in a depression, some of the water sinks into the soil. Infiltration is entry of water from the ground surface into the underlying earth material. The term infiltration contains the root of the word filter, meaning to pass through; the prefix signifies that the process is one of passing into. Infiltration rate is given in units of cm/h, the same type of units used to describe rainfall.

The amount of rainfall or surface water in excess of infiltration runs off the surface. Thus Horton (1933) showed (although by focusing too simplistically on surface processes) that more intense rain would cause more runoff from a given soil, because the excess quantity of water over infiltration would be greater, and that a soil with great ability to infiltrate water, such as sand or gravel, would create less runoff than one with less ability, such as unvegetated clay.

A well-vegetated upland soil surface can infiltrate water at surprisingly high rates. Despite the predictions of some early hydrologic models, under natural rainfall surface runoff is almost never generated on well-vegetated unsaturated soils. In wooded or thickly grassed areas of the humid eastern United States, plants and organic litter cover the soil, protecting it from the erosive impact of falling raindrops. Vast amounts of rainwater are conveyed into the soil by macropores, built partly by plant roots and stabilized partly by decomposed organic matter. By the time rainwater reaches purely mineral soil, the water is working under vadose processes in the soil matrix, and a large quantity of water has been stored in the surface layer for continued deeper percolation.

In contrast, runoff occurs frequently on many unvegetated upland surfaces such as soils denuded by construction, desert pavements and urban impervious surfaces. In such places the surface lacks a porous structure. Little infiltration occurs, and a large part of rainfall goes into direct surface runoff.

Infiltrated water moves downward into soil, partly because it is pulled from below by the matric suction of underlying soil that is drier than that near the surface. Movement into a soil stops when the wetting has progressed so far that all the water that has soaked into the soil is held by attraction to the grains and is in equilibrium with the matric suction of the surrounding soil.

If there is enough infiltration, some of the water draining downward through the vadose zone reaches the water table, raising the water table and enlarging the phreatic volume. In coves, lowlands and floodplains the phreatic zone discharges to streams, wetlands, lakes and oceans, completing the linkage from infiltration, through the vadose zone, to the phreatic zone, to discharge.

Along the way, water is lost from the subsurface by evapotranspiration. Some of this occurs as evaporation directly from wet soil surfaces into the atmosphere. The remainder is transpiration from plants. Plants withdraw available moisture from the soil root zone, reducing the amount available for ground water recharge and ultimately for stream flow.

The hydrologic balance

In any landscape, over any given time period, any difference between hydrologic inflow and outflow is resolved by a change in storage, in some combination of surface, vadose and phreatic forms. This is the hydrologic balance or water balance, ΔStorage = inflow minus outflow. For a simple drainage area, the water balance can be expanded into ΔStorage = P - Et - Ro, where P is precipitation, Et is evapotranspiration, and Ro is runoff. An analogous water balance can be constructed for any component of a watershed or landscape, such as the vadose zone or the phreatic zone, a development site, a parking lot or a political jurisdiction.

The water balance is a unifying theme in landscape hydrology. It is the site-specific expression of the hydrologic cycle, in which water circulates on a global scale from land, to ocean, to atmosphere, and back to land, with a temporary increase or decrease in storage balancing any difference between one end of the cycle and another (Ferguson, 1992). It is analogous to and linked with nutrient budgets, where water is a powerful solute, and energy budgets, where water has high latent heat. In general, places with similar water balances bear similar life forms with similar levels of biotic productivity and diversity (Meentemeyer and Elton, 1977; Walter, 1979; Wilson, 1992, p. 199-200). They have similar soil moisture processes (evaporation and accumulation of salts vs. leaching and ground water recharge), soil depth, drainage integration (playa lakes vs. integrated stream systems), quality of ground and surface waters, annual runoff, potential for local water supplies, and potential for increasing plant growth by irrigation.

When inflow exceeds outflow a water surplus exists, and storage increases. Increasing precipitation or decreasing evapotranspiration is reflected in greater soil moisture, reduced matric suction, and greater discharge from the vadose zone to the phreatic. The water table rises; the slope of the water table toward discharge points increases. Stream flow rises as ground water discharges more rapidly.

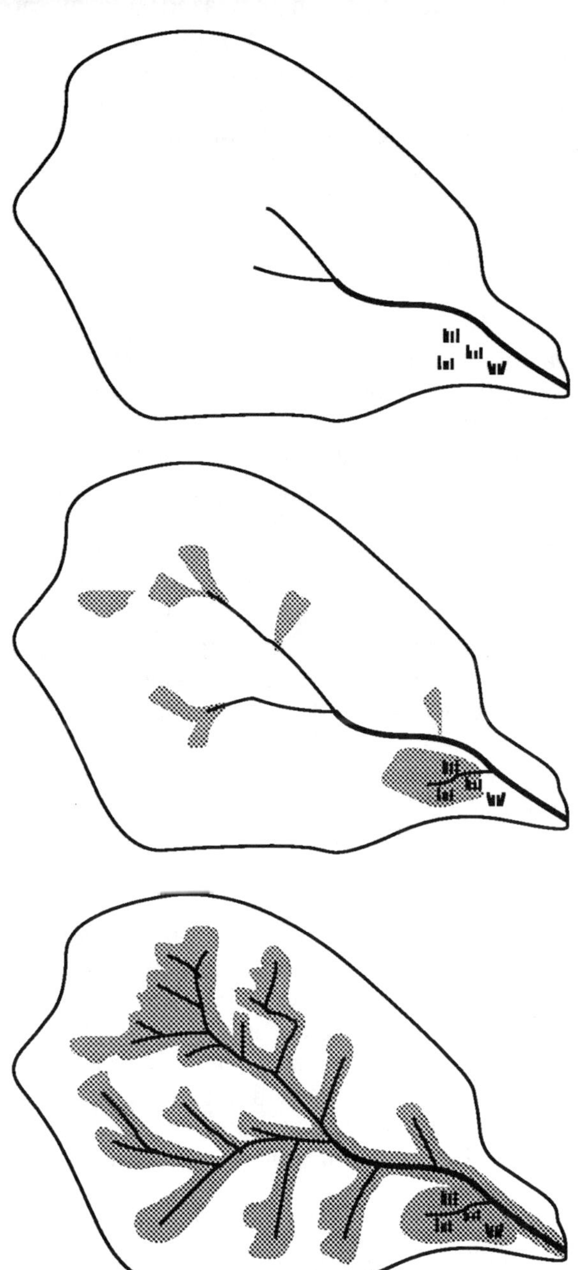

Figure 1.5 Growth of saturated soil areas in a watershed during a rainstorm. Saturated, runoff-producing zones are shown as shaded areas. The top figure shows the watershed immediately before the storm, the middle figure during the rise of the stream hydrograph, and the bottom figure at the peak of stream discharge (after Hewlett, 1982, Figure 7-12)

In wet periods, when infiltration and throughflow are active, the raised water table is reflected in an increased area of saturated soil in a watershed (Dunne, Moore and Taylor, 1975; Hewlett and Hibbert, 1967). Saturated areas can occur in stream valleys, swales, and other concave and low-lying slopes that intersect the water table or concentrate downslope vadose flow. Figure 1.5 shows how, in wet periods, the saturated areas expand out of the immediate stream sides into hillslope hollows. The saturated areas yield surface runoff by draining subsurface water to the surface and directly from rain falling on the saturated areas where infiltration is not possible.

When outflow exceeds inflow, a water deficit exists, and storage and throughflow decline. Subsurface water continues to flow toward streams while the mound of ground water gradually flattens out. As moisture drains out of storage zones without replenishment, saturated runoff-producing zones shrink. Evapotranspiration at and near the soil surface progressively sets up matric suction, interrupting downward drainage.

During a drying period discharge from a watershed does not cease abruptly, but rather declines more and more slowly. The lower the moisture content in the vadose zone, the greater the matric suction, and the lower the rate of drainage downward to the water table. The lower the gradient of the water table toward discharge points, the lower its rate of discharge to surface streams. Low storage levels bring about conservation of the remaining water. Low discharge rates may continue for months after the most recent rains.

Storm flows and base flows

Stream flow is characteristically variable, within limits defined by the drainage area and the climate. Figure 1.6 shows the daily discharge in Atlanta's Peachtree Creek for four months in the spring and early summer of 1960. The abrupt upward spikes of storm flows are easily distinguished from the base flow, the gently undulating platform or "base" to which flow returns following storm events. The total volume of water that discharges from the watershed over the course of a year is a composite of storm flows and base flows.

Storm flows tend to reflect more or less direct surface runoff from rainfall or snowmelt, particularly from saturated runoff-producing zones and urban impervious surfaces. Surface runoff

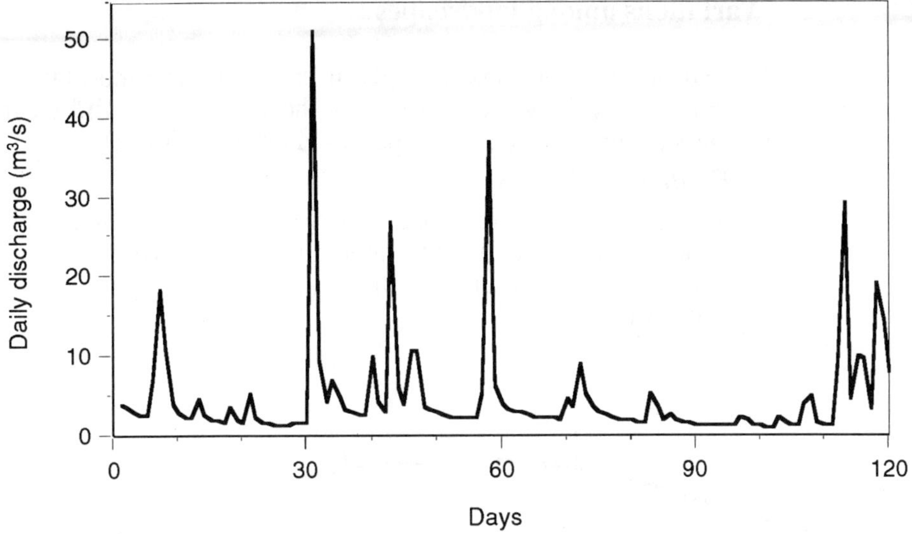

Figure 1.6 Daily discharge of Peachtree Creek, Atlanta, Georgia, from March through June 1960 (data from U.S. Geological Survey).

can reach a stream in minutes. The receding limb of each storm flow merges gradually into the general base flow as subsurface processes take over from surface ones.

The base flow undulates seasonally with the water balance accumulated over the previous several months. Peachtree Creek's base flow is relatively high in the late winter and spring, following winter's low evapotranspiration, sloping gradually to lower levels in summer as evapotranspiration and continued drainage compete with precipitation for dominance of the water balance. Base flow reflects the gradual discharge of water that infiltrated into the subsurface and was held in slow-moving vadose and phreatic storage long after any individual storm event. Rainwater emerges from subsurface storage so gradually that its discharge is part of the streams' base flow, not the original storm event. The flow is qualitatively different. It has become a different type of flow, supporting different types of resources.

Variations among landscapes

The hydrologic functions of a given landscape are maintained at a general level by the landform—the topography and earth materials through which water passes (Ferguson, 1992).

Positive landforms

Positive landforms are those that rise above the water table and major streams, such that water ultimately moves outward from the landform. Figure 1.7 distinguishes four types of positive landforms:

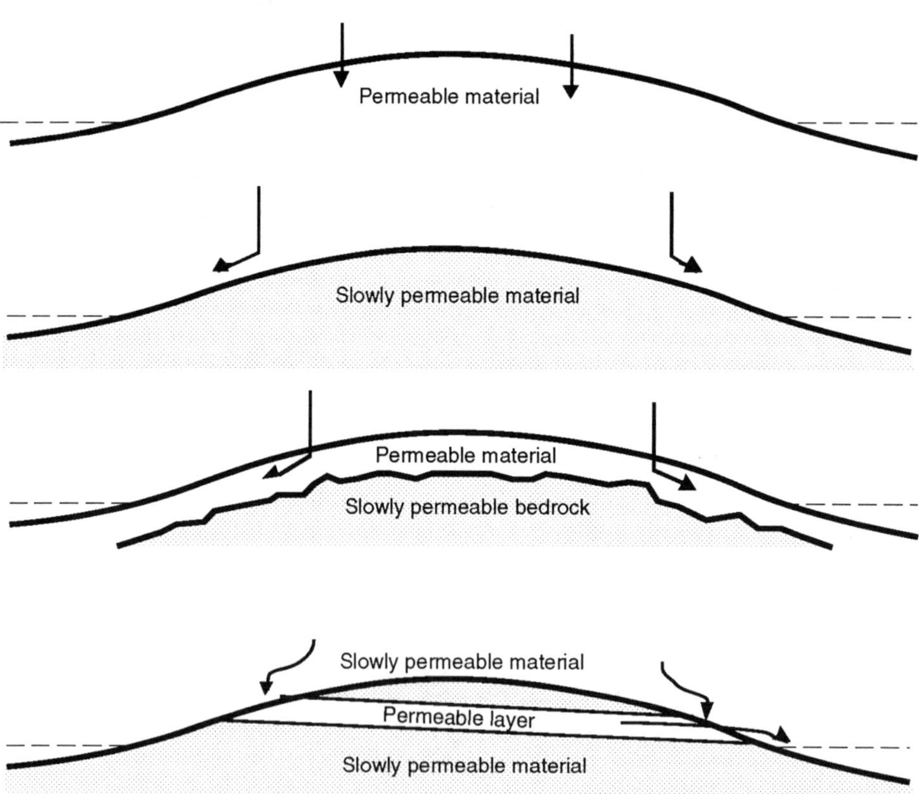

Figure 1.7 Positive landforms: from top to bottom, the landforms are water infiltrating, water spreading, hybrid, and layered. The dashed lines represent the elevation of the water table and major streams.

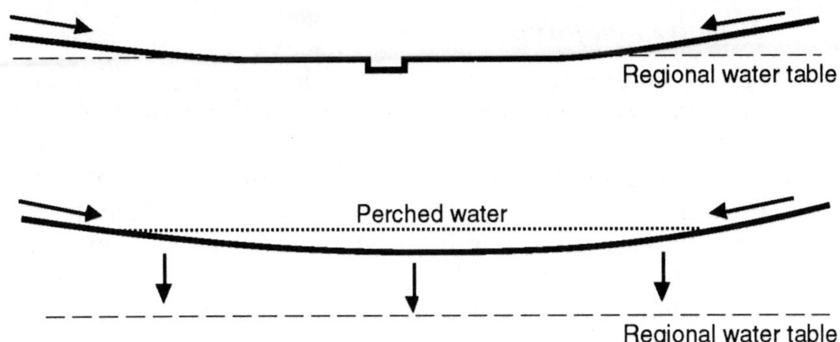

Figure 1.8 Negative landforms: continuous in the upper figure, perched in the lower. The dashed lines represent the elevation of the water table and major streams.

• Water-infiltrating landforms are those with entirely permeable bodies. Examples are sandstone, carbonates, and most unconsolidated materials. A continuous mass of ground water exists under the entire landform. Infiltrated water that does not reach the water table is stored in and flows through the large vadose zone.

• Water-spreading landforms are essentially impermeable throughout their depth, including impermeable bedrock with little overlying soil. Examples are many occurrences of shale, slate and granite. They are familiar in arid regions, where soils may be shallow to absent. Infiltration and subsurface storages and flows are small. Drainage is mostly by surface runoff.

• Hybrid landforms have essentially impermeable, water-spreading bedrock, but a significant mantle of permeable soil. An example is the deeply weathered crystalline rock of the southern Piedmont. Water infiltrates the saprolite, forms a film of phreatic water over the bedrock surface, and accumulates in rock fractures and stream valleys. By default, vadose moisture is a major part of total subsurface storages and flows.

• Layered landforms have a relatively impermeable surface layer, underlain by a more permeable zone. An anthropogenic example is an impermeable urban surface over a permeable soil mantle. Some sedimentary plateaus can exhibit many successive layers of materials, permeable and impermeable, between the land surface and the base level of major streams.

Negative landforms

Negative landforms are those at or below the water table and the elevations of major streams, such that water moves inward to the landform, creating at least intermittent surface water flows and storages. During times of water surplus they receive the discharge from adjacent positive landforms, often transferring it down regional drainage gradients. Vadose flows and storages are relatively insignificant; essentially all subsurface flow is saturated. Figure 1.8 distinguishes two types of negative landforms:

- Continuous landforms are those hydraulically in contact with regional ground water or streams. They are characteristically shaped by geologically recent fluvial processes. They lace through all regions in the forms of floodplains and coastal marshes.
- Perched landforms are those resting on slowly permeable material above regional ground water and stream levels. Examples are many playas and vernal pools, many lakes and wetlands on glacial till, many karst sinkhole lakes, hillside springs in dissected sedimentary plateaus, streams discharging from mountain canyons onto alluvial fans above the valley's water table, and anthropogenically lined pools. During times of water surplus they may overflow into surrounding permeable materials or ephemeral surface channels.

Urban impacts

Urbanization involves the sealing of the land surface with pavements and roofs, creating a hydrologically new kind of landform.

Urban impervious surfaces alter all parts of the hydrologic balance (Lerner, 1990). They deflect rain water away from infiltration, soil moisture, recharge, subsurface storage and base flow. Impervious surfaces prevent water from entering the natural structures and processes that maintain hydrologic storages and moderate hydrologic flows.

The deflection away from subsurface paths creates runoff that moves quickly down stream systems and out of the landscapes where it originated. This effect is reflected in storm flow rate and volume. Leopold (1968) found for the Brandywine

River that the frequency of storm flow of a given magnitude increased with urbanization of the watershed. Ferguson and Suckling (1990) found for Peachtree Creek that for a given amount of storm precipitation the flow volume increased with urbanization. Stream erosion follows the altered storm flow rate and duration (McCuen and Moglen, 1988), making stream water turbid even where excess sediment is not delivered from the uplands.

Detention basins required by local governments for new developments fail to stop the increase. Detention basins aggravate flooding as often as they reduce it, because they allow all the water that flows into them to flow on into streams. The flows from different tributaries combine; with total volumes of flow unchanged, there is no possibility of escaping the mounting rates of flow downstream. Even where detention successfully suppresses peak flow rate, it does not prevent overflows of mounting volumes backed up at culverts, fills and other common urban drainage obstructions (Ferguson and Deak, in press). Detention addresses some of the symptoms of urbanization that appear in the stream system, but the cause of the stormwater problem remains in the upland soil where natural infiltration has been bypassed.

Urbanization's deflection of flows away from subsurface paths makes base flows decline. During urbanization of Peachtree Creek's watershed, annual low flow decreased for a given amount of prior precipitation, particularly during dry years. Declining base flows are environmentally and economically critical: base flows must be sufficient to absorb pollution from sewage treatment plants and non-point sources, support aquatic life dependent on stream flow, and replenish water-supply reservoirs for municipal use in the seasons when lake levels tend to be lowest and water demands highest. Declining base flows reflect declining soil moisture and ground water levels. If the water is no longer in the watershed, having been shunted away by impervious surfaces during storm events, then there is no possibility of fully restoring base flows through reservoirs or any other kind of surface controls.

STORMWATER AS A RESOURCE

Resource for ground water recharge

On all sides of a water supply well, ground water drains toward the suction created by the pump. The well creates a cone of depression in the water table or piezometric surface, with its low point at the well. When water is pumped rapidly, the cone of depression is deeper and steeper than when the pumping is slow, and water moves correspondingly faster toward the well.

The flux of water required to support modern cities is greater than that of any other substance (Wolman, 1965). As a resource, water is limited in renewability and degradable in quality.

When pumpage and natural drainage together exceed an aquifer's rate of inflow, then the volume in storage is drawn down, and the water table or piezometric surface falls. The energy cost of pumping increases in proportion to the depth from which water is lifted. The relaxed pressure allows saline water from deeper aquifers or a nearby seacoast to contaminate the aquifer.

In some areas, precipitation is so scanty that only occasionally does enough fall to add any appreciable amount to the phreatic zone, yet wells encounter great masses of ground water. In such areas water is being pumped that accumulated under a different climatic regime. This is fossil water, in the sense that it was deposited in earlier geologic ages; to pump it out is to mine it like any other nonrenewable resource. Such water use is clearly not sustainable, as water that took thousands of years to deposit is being mined at a rate that will exhaust it in relatively few years.

Even in humid areas where ground water is regularly replenished under current climatic regimes, rates of anthropogenic withdrawal can exceed the recharge rate. Such overdrafts are as unsustainable as those in arid regions.

If pumping ceases, inflow might sooner or later replace the water withdrawn, and the water table or piezometric surface return to a balanced, stable level. Rebounding may take months or many years, depending on the aquifer's recharge rate.

Thus for a given aquifer a finite sustainable ground water yield can be defined based on rate of recharge. Quantitative determination of replenishment rate is difficult for many aquifers, but the principal of finite supply applies to all aquifers.

Ground water replenishment can be maintained and supplemented and sustainable yield increased by artificial recharge and stormwater infiltration. The maintenance of the recharge function through natural recharge zones that would otherwise be paved over is essential for upholding the amount and quality of ground water. Replenishment during wet periods when surplus water is available results in subsurface storage for possible withdrawal during dry periods.

Resource for stream flow

Directions of flow between subsurface and surface water bodies correspond with types of negative landforms.

The most common is ground water discharge, which occurs in continuous landforms such as perennial streams, certain playa edges, and boggy slopes on glacial till (Siegel, 1983). Ground water slopes toward and discharges to the surface water body. The linkage between subsurface and surface waters is illustrate by the fact that, in some areas, high rates of ground water pumpage have resulted in loss of discharge to nearby surface water bodies.

Ground water recharge occurs in perched landforms. Water drains from the surface to the water table. Some perched lakes in Florida are known to periodically disappear when underground drainage is unplugged by erosion through underlying fissures (Lane, 1986).

In some places flow direction reverses seasonally between discharge and recharge. Examples are many floodplains of the southern coastal plain and the Mississippi valley, where the flow of springs reverses to reflect relative elevations of the stream and the adjacent water table.

Despite the variability of stream flow, it is practically uniform when compared with the discrete, abrupt, intense and often widely separated precipitation inflows that created it (Hewlett and Hibbert, 1967). Infiltrated water is not available as direct surface runoff. By damping surges of direct runoff, infiltration protects the stream and its ecosystem from frequent flushing and erosion. The only known way to suppress storm flow volume is stormwater infiltration. Its effectiveness in controlling volume does not depend on location of basins within a watershed. As a by-product of reducing volume, infiltration also tends to reduce peak flow rate (Ellington and Ferguson, 1991).

Many surface streams like Peachtree Creek, where the water table is high enough to discharge to the surface drainage system, can continue to flow during long periods of dry weather. Water infiltrated into and stored in the subsurface emerges during later base-flow periods. Perennial stream flow and the volume of water in stream pools during low-flow periods determine a stream's stability and productivity for aquatic ecosystems, human water supply, recreation and aesthetic interpretation. Stream water quality is a function of the stream's dilution capacity (quantity of base flow) relative to the waste load from the drainage basin.

Thus infiltration on upland surfaces, seemingly distant from streams, plays a decisive role in stream flow regime and the protection of aquatic resources.

Resource for biota

Transpiration—the vaporization of water from plant cells—pulls water from the soil into roots and up plant stems, raising soil nutrients into plants and supplying moisture to protoplasm. The graph in Figure 1.9 shows that productivity of plant communities is directly proportional to evapotranspiration rate.

Meteorologic evapotranspirative potential, called reference evapotranspiration or potential evapotranspiration, exerts a daily pull on moisture from plant stomata. When plant rootlets transfer the pull by absorbing water from soil, moisture films in the vicinity are thinned and their matric suction is increased. Water tends to obey the suction gradient by moving toward the points of root absorption. The rate of movement depends on the conductivity of the soil and the magnitude of the suction gradient. Plant protoplasm dies when its water content falls below 30 to 50 percent (Daubenmire, 1959, p. 130-136). Given the varying level of moisture in the soil, plants are presented with the problem of maintaining a satisfactory balance between loss and absorption of water.

The life cycles of most organisms are organized around their access to water. Through feedback loops, biological processes may modify the hydrologic setting (National Research Council,

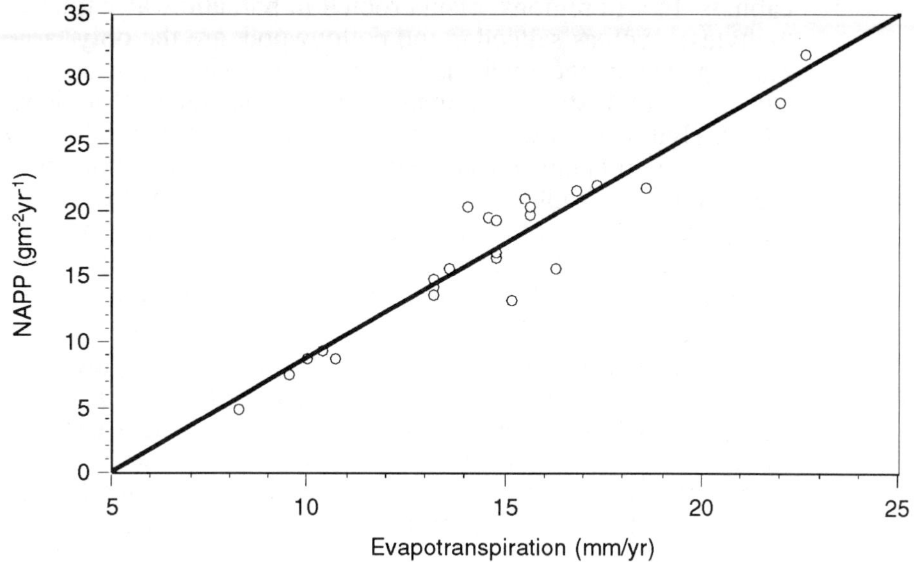

Figure 1.9 Relationship between net above-ground primary productivity of plant communities (NAPP) and evapotranspiration rate (data from Rosenzweig, 1968).

1991, p. 167-178). In 1895 E. Warming distinguished classes of plants in terms of their typical environments: xerophyte, mesophyte and hydrophyte, combining the Greek phyton, meaning plant, with references to dry, intermediate and wet habitats, respectively.

Xerophytes inhabit deserts and dry ridge tops. They have adapted to low soil moisture levels with ephemeral life cycles, succulence, coverings of nearly impervious tissue that reduce transpiration, extensively branched roots that extract moisture from the soil aggressively, guard cells that regulate stomatal size, and reduction in sizes of leaves and cells.

At the other extreme, hydrophytes inhabit wetlands and stream corridors (Mitsch and Gosselink, 1993, p. 507-527). The productivity of wetlands can be the highest of all natural communities in a given region. About half of all endangered and threatened animals species, and more than one third of the endangered and threatened plant species, depend on wetland habitats. In arid regions, plants rooted in phreatic water (phreatophytes), such as saltcedar and cottonwood, are the only large green plants in the open desert landscape, signalling springs and ephemeral stream channels from a distance (Robinson, 1958). Hydric soils tend to be anoxic, altering the availability of nutrients to plants and preventing the aerobic metabolism of roots. Many hydrophytes have adapted by development of aerenchyma (air spaces), diffusing oxygen to flooded roots and stems. Others have adapted with timing of seeding in the non-flood season and with altered photosynthetic chemistry.

In an established plant community, when ground water rises due to blocked drainage or inflow of excess stormwater, the water displaces oxygen in soil pores, calling forth mechanisms to supply oxygen to roots, or for the roots to use less oxygen: extension of roots at more oxygenated levels in the soil, development of aerenchyma tissue, mortality of unadaptable individuals and species, and replacement by hydric species.

When ground water and soil moisture decline due to interruption of infiltration and recharge, it calls forth a sequence of mechanisms that develop deeper water sources or conserve water by reducing evapotranspiration: stomatal closure and root extension by adaptable organisms, dormancy and death of an increasing number of individuals and species, and replacement by more xeric species.

Infiltration of rainwater and runoff near the source can maintain moisture in nearby soils at predevelopment levels. It can avoid surface diversions that increase water levels in some spots while reducing it in others. It can prevent biotic stress in the surroundings and downstream.

Maintenance of landscape hydrology

A landscape is a three-dimensional mosaic of interacting environmental compartments or zones. A specified locale can be considered, at a variety of scales in space, as a coherent entity where structure, function and development can be delineated.

Hydrology interacts with other components such as air, nutrients, soil and biota to form landscape patterns with implications for habitat and sustainability (Ferguson, 1992). Hydrologic form and process interact to assure a dynamic equilibrium in which the system has the flexibility to adjust through changing circumstance. Subsurface water occupies a unique and essential place in the environmental functions of landscapes.

Man modifies and interacts with natural flows and storages of water in a manner analogous to agriculture's modification of and interaction with "natural" soils and plants. Mankind's water management provides stimuli to which environmental compartments respond according to their native properties.

Mankind's clearing, paving, damming and channelization have often been done with the intent of subduing and controlling nature, but instead have bound nature's dynamic mechanisms into distorted cycles of erosion, extinction and stress. While accommodating human use, the natural operation of environmental adjustment mechanisms should not be restricted. The landscape must be supported in its battle against stress. If our sense of the hydrologic continuum is fragmented, then our water management will follow in the same path. Preservation of the integrity of the continuum ought to be one of the objectives of resource use.

Stormwater management is not a mechanical system or a utility. Stormwater is a resource. Its management is an environmental process, joining the atmosphere, the soil, vegetation, land use and stream flows.

Our understanding of urban hydrology must encompass vegetation and evapotranspiration as well as impervious surfaces and direct runoff. Our conception of stormwater management must encompass base flows as well as peak flows. It needs to address long-term storage in subsurface voids as well as the visible dynamism of surface flows. It needs to reinitiate the kinds of long-term environmental processes that occurred before impervious surfaces were installed.

Stormwater infiltration addresses the cause of the urban stormwater problem at the soil surface where development takes place. Infiltration controls the volume of storm runoff, keeping aggravated storm surges out of streams and returning the flow volume to its place in the soil and long-term base flows. It uses the natural capacities of soil, vegetation and landforms for restoring the direction, timing and quality of hydrologic flows. It promises to restore the hydrologic balance of urban landscapes, returning hydrologic storages and flow regimes, and the ecosystems of which they are part, to a self-sustaining equilibrium.

REFERENCES

Adams, F.D., 1938, *The Birth and Development of the Geological Sciences*, New York: Dover.

Baver, L.D., Walter H. Gardner and Wilford R. Gardner, 1972, *Soil Physics*, fourth edition, New York: Wiley.

Beven, Keith, and Peter Germann, 1982, Macropores and Water Flow in Soils, *Water Resources Research* vol. 18, no. 5, pages 1311-1325.

Brady, Nyle C., 1974, *The Nature and Properties of Soils*, eighth edition, New York: Macmillan.

Daubenmire, R.F., 1959, *Plants and Environment*, second edition, New York: Wiley.

Dunne, T., T.R. Moore and C.H. Taylor, 1975, Recognition and Prediction of Runoff-Producing Zones in Humid Regions, *Hydrological Sciences Bulletin* vol. 20, pages 305-327.

Ellington, M. Morgan, and Bruce K. Ferguson, 1991, Comparison of Infiltration and Detention in the Georgia Piedmont Using Recent Hydrologic Models, pages 213-216 of *Proceedings of the 1991 Georgia Water Resources Conference*, Kathryn Hatcher, editor, Athens: University of Georgia Institute of Natural Resources.

Ferguson, Bruce K., 1992, Landscape Hydrology, A Component of Landscape Ecology, *Journal of Environmental Systems* vol. 21, no. 3, pages 193-205.

Ferguson, Bruce K., and Tamas Deak, (in press), The Role of Urban Storm Flow Volume in Local Drainage Problems, accepted for publication in *Water Resources Planning and Management.*

Ferguson, Bruce K. and Philip W. Suckling, 1990, Changing Rainfall-Runoff Relationships in the Urbanizing Peachtree Creek Watershed, Atlanta, Georgia, *Water Resources Bulletin* vol. 26, no. 2, pages 313-322.

Heath, Ralph C., 1983, *Basic Ground-Water Hydrology*, Water-Supply Paper 2220, Washington, D.C.: U.S. Geological Survey.

Hewlett, J.D., 1982, *Principles of Forest Hydrology*, Athens: University of Georgia Press.

Hewlett, J.D., and A.R. Hibbert, 1967, Factors Affecting the Response of Small Watersheds to Precipitation in Humid Areas, pages 275-290 of *International Symposium on Forest Hydrology*, W.E. Sopper and H.W. Lull, editors, Oxford: Pergamon.

Horton, Robert E., 1933, The Role of Infiltration in the Hydrologic Cycle, *Transactions of the American Geophysical Union* vol. 14, pages 446-460.

Jury, William A., Wilford R. Gardner and Walter H. Gardner, 1991, *Soil Physics*, fifth edition, New York: Wiley.

Lane, Ed, 1986, *Karst in Florida*, Special Publication No. 29, Tallahassee: Florida Geologic Survey.

Lerner, David N., 1990, Groundwater Recharge in Urban Areas, *Atmospheric Environment* vol. 24B, no. 1, pages 29-33.

Leopold, Luna B., 1968, *Hydrology for Urban Land Planning, A Guidebook on the Hydrologic Effects of Urban Land Use*, Circular 554, Washington, D.C.: U.S. Geological Survey.

Leopold, Luna B., 1974, *Water, A Primer*, San Francisco: Freeman.

McCuen, Richard H., and Glenn E. Moglen, 1988, Multicriterion Stormwater Management Methods, *Journal of Water Resources Planning and Management* vol. 114, no. 4, pages 414-431.

Meentemeyer, V., and W. Elton, 1977, The Potential Implementation of Biogeochemical Cycles in Biogeography, *The Professional Geographer* vol. 29, no. 3, pages 266-271.

Mitsch, William J., and James G. Gosselink, 1993, *Wetlands*, second edition, New York: Van Nostrand Reinhold.

National Research Council, 1991, *Opportunities in the Hydrologic Sciences*, Washington, D.C.: National Academy Press.

Robinson, T.W., 1958, *Phreatophytes*, Water-Supply Paper 1423, Washington, D.C.: U.S. Geological Survey.

Rosenzweig, Michael L., 1968, Net Primary Productivity of Terrestrial Communities: Prediction from Climatological Data, *The American Naturalist* vol. 102, no. 923, pages 67-74.

Schultz, Richard C., and John D. Hewlett, 1978, Soil Moisture as a Part of the Hydrologic Cycle, pages 7-21 of *Proceedings, Soil Moisture...Site Productivity Symposium*, U.S. Forest Service, Myrtle Beach, S.C., Nov. 1-3, 1977.

Siegel, D.I., 1983, Ground Water and the Evolution of Patterned Mires, Glacial Lake Agassiz Peatlands, Northern Minnesota, *Journal of Ecology* vol. 71, pages 913-921.

van der Leeden, Frits, Fred L. Troise and David Keith Todd, 1990, *The Water Encyclopedia*, second edition, Chelsea, MI: Lewis.

Walter, H., 1979, *Vegetation of the Earth and Ecological Systems of the Geobiosphere*, second edition, New York: Springer-Verlag.

Wilson, Edward O., 1992, *The Diversity of Life*, New York: Norton.

Wolman, A., 1965, The Metabolism of Cities, *Scientific American* vol. 213, pages 179-190.

2

Infiltration Layout and Construction

Stormwater infiltration is executed in pervious areas of the land surface, and in surface and subsurface infiltration basins.

Any approach to layout and construction of infiltration surfaces and basins must be flexible, creative and site-specific. Each surface or basin occupies a functional and perceptual space in a site, having at least general requirements for dimensions, landform, vegetation and relationships to surrounding land uses. Safety, maintenance, aesthetics, human use and environmental equilibrium should be concerns for infiltration surfaces and basins, as they are for the streets, shopping centers, work places and homes that they serve (Ferguson, 1991). Within a few hydraulically derived parameters for volume and dimensions, a surface's or basin's materials and plantings can be selected, and its contours can be molded. The broad views and intuitive art of urban and environmental design must be considered simultaneously with mathematical derivation for hydraulic performance. The solutions are most successful when considered at the earliest stages of site planning.

Infiltration can be effectively implemented in numerous infiltration surfaces and small basins near the sources of direct runoff, leaving low-lying wetlands untouched and restoring the hydrologic function of the uplands to the watershed as a whole. As an alternative, surface runoff can be conveyed downstream to a point in the watershed where infiltration is uniquely feasible and desirable, such as the linear aquifer recharge zone in Los Angeles where spreading basins are installed.

Every drainage structure has both primary and secondary components (Jones, 1967; Newville, 1967). The primary system is designed to meet hydraulic objectives at all times up to the occurrence of a given design storm. The secondary system, where deliberately designed, is intended to prevent overflowing water from causing damage when the primary system is overloaded or inoperative. Every inlet in every street in the world overflows when a storm larger than the design storm occurs or when the inlet is clogged; at that time the secondary system goes into operation, and water flows down the gutter, across the pavement, or over the curb downslope. This type of behavior occurs in drainage systems of all types, whether it was planned for or not. Infiltration surfaces and basins behave the same when they, too, experience overly wet periods or are clogged: the surface or basin overflows, and the excess water flows overland. Where the water will go across the surface at that time must be planned. A large basin must include a nonerodible spillway or bypass to confine large overflows to planned surface drainage routes. The overflow route must be out of the way of buildings, concentrated traffic, and other places that could be seriously damaged by excess water.

Configurations of slopes, soils, site boundaries, subwatersheds and land use intensities sometimes prevent infiltration from being fully implemented. Hybrid plans may be needed, in which some runoff, or the runoff from only part of the site, is infiltrated, and the rest is treated by artificial wetlands or other surface measures.

INFILTRATION SURFACES

All pervious areas in a site are resources for maintaining and enhancing infiltration. Where the volume of runoff is reduced at the source by infiltration through porous surfaces, less land or construction effort is needed for downstream basins to control the remaining runoff.

Where the soil must be paved to stabilize the surface for traffic, pavements can be made of permeable materials. Where impervious surfaces are required, grading can direct their discharge to nonerodible pervious surfaces, where at least some water has a chance to infiltrate. Even where roofed buildings are required, pervious surfaces can be partly reclaimed by constructing the buildings underground (Wells, 1981).

A porous surface can be designed to meet a variety of hydrologic objectives: to infiltrate all rainwater that falls on it, to infiltrate enough precipitation that runoff is not increased by development of the surface, or to infiltrate both direct precipitation and runoff from adjacent developed areas. Thus a surface such as a parking lot that has the potential to create a great deal of surface runoff could be designed to produce no runoff, or no more than occurred prior to development.

Spreading out infiltrating water over the largest possible pervious area eliminates the largest possible runoff volume. Maximizing vegetated areas, porous pavements and other infiltration surfaces is particularly beneficial on sites with slowly permeable soil. As detailed in Chapters 3 and 4, low soil permeability makes spreading out necessary in order to infiltrate a given volume of water within an acceptable time period.

Vegetated surfaces

Vegetation and accompanying fauna maintain soil aggregation, macropores and infiltration rates. The highest infiltration rates in a given soil type are found in maturely vegetated areas free of clearing, grading, compaction and disturbance. Site layout should seek to preserve existing vegetated areas, to restore successional vegetation on parts of sites disturbed by construction or prior land use, and to minimize paved areas.

The area of urban lawns in the United States is bigger than that of any agricultural crop (Bormann, Balmori and Geballe, 1993, p. 68). Hamilton (1990) measured infiltration into home lawns and turf plots on fine- and medium-textured soils in Pennsylvania. The grasses were various mixtures of Kentucky bluegrass (*Poa pratensis*), perennial ryegrass (*Lolium perenne*) and fine fescue (*Festuca* spp.). The results varied greatly, from 0.4 to 10.0 cm/h. The lowest rates were on the most recently established lawns (2 years old), where macropores were little developed. The highest rates were on the oldest lawns and on a

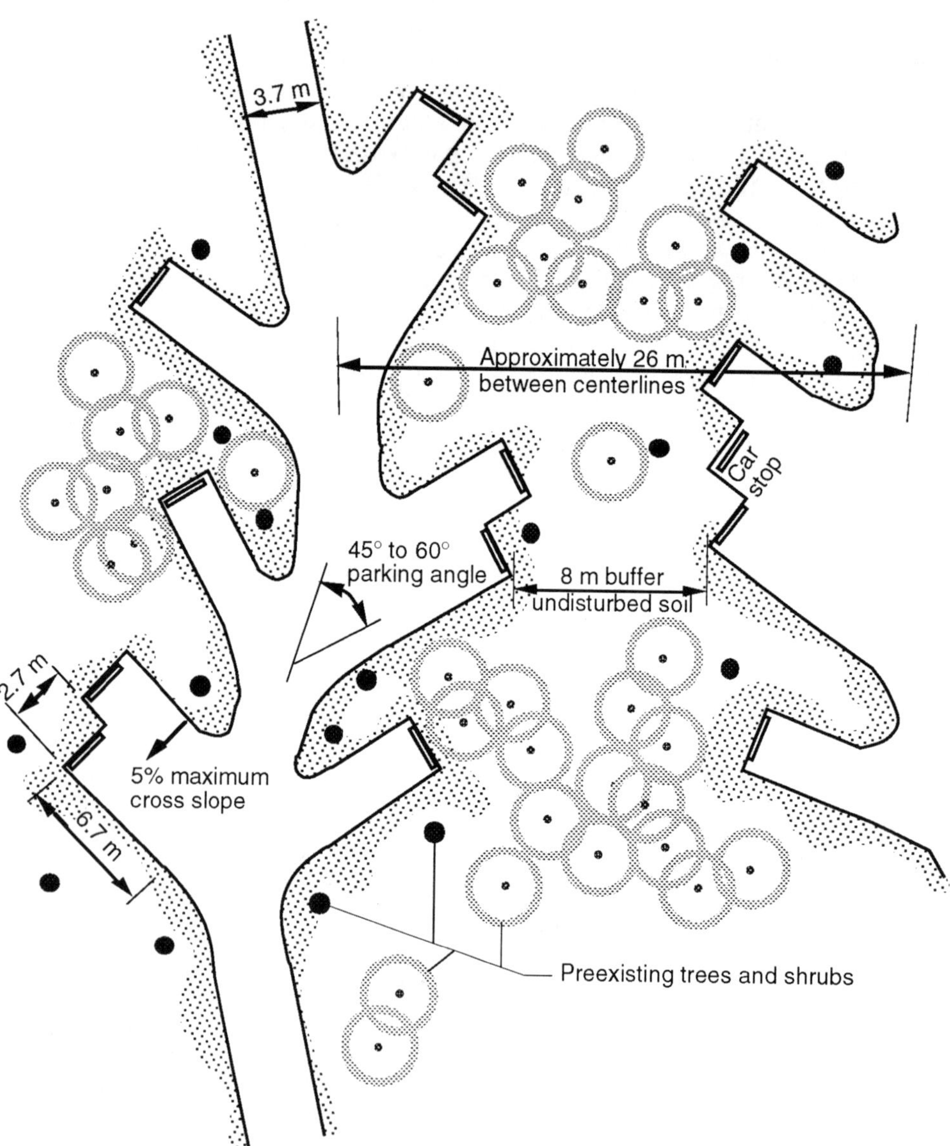

Figure 2.1 Woodland parking at the Jones Bridge headquarters of the Simmons Company (after Marvin and Paddock, 1979).

lawn area not excavated during any part of construction. Infiltration rates were limited by the soil's structural condition following earthwork and site preparation, no matter how dense or healthy the turf afterwards. The physical conditions of a lawn soil are created by soil handling during construction, including topsoil stripping, traffic on exposed subsoil, and stratification of soil upon replacement. To achieve high infiltration rates on lawns and other areas around construction, construction must be overseen vigilantly to minimize earthwork, preserve soil structure and avoid compaction. In addition lawn management needs to encourage earthworms for rapid macropore redevelopment. As detailed in Chapter 6, adding gypsum to a disturbed soil surface can encourage development of a porous, permeable soil structure.

A project at the Riverbend East residential community in Athens, Georgia managed to restore the infiltration capacity of a 10-year-old compacted, poorly vegetated lawn. In 1983 the soil in sloping residential side yards was eroding under runoff from an upslope parking lot. The bare, eroded, slowly permeable soil surface multiplied the runoff volume. The runoff was dissipated by dividing the soil into 10 m^2 terraces using timber railroad ties slightly embedded in the soil. A slight cross-slope on the ties deflected runoff from each terrace into an adjacent wooded area for infiltration in the native vegetated soil. Ten years later, the terraces were vegetated with grass and were highly pervious; concentrated surface runoff was not being passed downslope.

An analogous idea is employed in Germany, where arrays of "gravel bars" are placed across compacted and eroding slopes. Each bar is a narrow (0.2 m) trench filled with stone or gravel, a meter or two long and a half meter deep. Each bar traps the runoff from its miniature drainage area. The surrounding ground is loosened and seeded. Eventually spreading grasses root in the gravel and cover the gravel surfaces.

In the Atlanta area, at the 28-ha Jones Bridge headquarters of the Simmons Company, Robert E. Marvin and Associates (1978; Marvin and Paddock, 1979) aggressively protected pervious forest soil in the midst of industrial development. They broke 200 parking spaces into groups of two and three and dispersed them among preexisting pine and hardwood trees. Figure 2.1 shows the pattern of parking pavements that resulted. Following site planning, red surveyor's ribbon was staked around the building zone, along both sides of roadways, and around each parking space so the designers could see the layout on the ground, allowing adjustments to be made for trees, topography and drainage. Only the trees within the adjusted ribboned zone were allowed to be removed. In the parking areas, one-way lanes were cleared 26 m apart, leaving 8 m buffers of undisturbed soil and vegetation between parking spaces to absorb runoff and maintain a vegetative canopy over the entire area. The roads and parking were stabilized with stone before building construction began, providing space for construction staging without further site disturbance. The 140-m-long building was raised off the ground on trusses to allow uninterrupted surface and subsurface drainage and saving of trees within a few meters of the structure. Disturbed areas were revegetated with native species throughout the site. No drainage conveyance structures were installed except for stone to stabilize downspout discharge points and drainage swales under the building. After 20 years, the condition of the site still shows that the pervasive undisturbed forest floor has largely absorbed the small, dispersed runoff volumes produced by pavements and downspouts, despite the fine texture of the native soil. Infiltration could have been still more successful if pavements had been made of any of the pervious materials which were just becoming available as this project was being completed.

At the Monarch apartment site on Hilton Head Island, South Carolina, Marvin used a similar approach on a level, sandy site among preexisting pines and palmettos. In this case there was no artificial paving on the individual parking spaces; cars parked directly on the highly pervious native sand.

Permeable pavements

Permeable pavements include porous asphalt and concrete, open-celled pavers, and a variety of other materials. Most permeable paving materials can be used in parking lots, lightly traveled streets, and pedestrian ways—in other words, most of the paved surfaces that are installed in urban communities. They can be used on subgrade slopes up to a few percent without subgrade erosion, but laying out a site from the beginning to create level slopes for their placement can maximize infiltration and make subgrade erosion impossible.

Considering pavement costs alone, some porous pavements have installation and maintenance costs approximately 10 percent greater than those of other types of pavements (Nichols, 1992; Sorvig, 1993). However, porous paving can be 12 to 38 percent less expensive when a site development is considered as a whole, because storm drainage systems can be significantly shorter and smaller where runoff is not generated (Sorvig, 1993).

The structural design of a porous pavement follows the same principles as the design of any other pavement. The bearing value of the soil should be determined by lab test or categorically by identification of soil type. Pavement thickness should be sufficient to carry the anticipated traffic load and meet frost conditions. The structural requirements for the soil under permeable pavements are the same as those for any other pavement and are easily met on many urban development sites: they must be permeable to water, must not heave during freeze-thaw cycles, and must not swell or turn liquid when wet.

Some permeable pavements have been fitted with drains to pass storm flows out to receiving streams after temporary storage in the pavement's base course, without forcing it to infiltrate the soil. These are detention systems that control only the rate of surface flow; they are not infiltration systems.

Porous asphalt

Porous asphalt consists of an open-graded asphalt concrete over an open-graded aggregate base, over a draining soil. Thelen and Howe (1978) presented the cross section shown in Figure 2.2 to illustrate the concept.

Open-graded asphalt concrete, like other asphalt concretes, is composed of stone aggregate and an asphalt binder (Diniz, 1980). It differs from other asphalt concretes in that it contains very little fine aggregate (dust or sand). Without fines filling the voids between the larger particles, the material is porous and permeable. A minimum of 12 percent void space by volume is frequently specified (Girling, 1992), ending up in the field typically around 16 percent, as compared with 2 to 3 percent in other asphalt concretes (Thelen and Howe, 1978, p. 9).

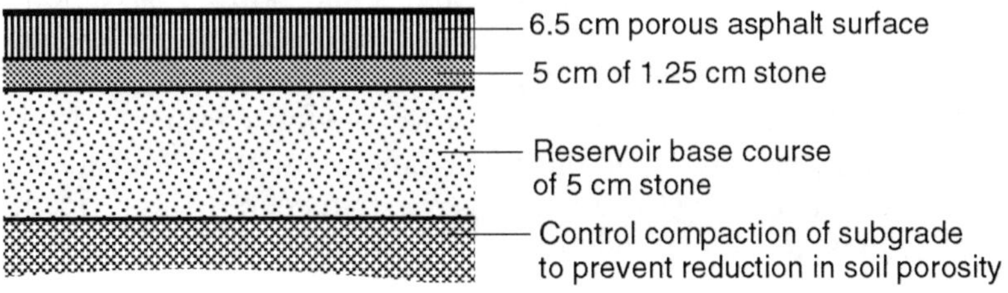

Figure 2.2 Illustrative cross section of a porous asphalt pavement (after Thelen and Howe, 1978, p. 4).

Open-graded asphalt is installed as a hot mix (Nichols, 1992). Asphalt is typically 5.5 to 6 percent by weight of the paving mixture. The lower limit is to create an adequately thick layer of asphalt around the aggregate particles; the upper limit is to prevent the mix from leaking asphalt during transport (Bachtle, 1974). Many asphalt plants routinely carry open-graded asphalt mixes, using names such as popcorn mix or porous friction coat (Sorvig, 1993).

The base material is open-graded crushed stone. The stone's void space of 38 to 40 percent is free for storage of water while the water percolates into the soil. The thickness of the base course can be adjusted to provide any needed hydraulic capacity. The top of the base course is choked with stones of intermediate size before placing the asphalt surface course in order to prevent collapse of the hot asphalt mix into the large voids of the base course.

Structurally, the thickness of asphalt surface and base courses used in conventional local practice should be equally satisfactory for porous asphalt carrying equivalent loads. Strength of open-graded asphalt comes from the friction and obstruction of the large angular particles trying to slide over one another (Bachtle, 1974). Porous asphalt is similar in Marshall properties (strength and flow) to other asphalt concretes. Deflection under a load is less than that in other asphalt pavements (Thelen and Howe, 1978, p. 13-14). It is equal in durability.

Porous asphalt pavement was conceived at the Franklin Institute in Philadelphia in 1968 and was developed there with EPA support in the early 1970s. The first field installation of porous asphalt was a driveway constructed in Delaware in 1973, underlain by a subbase of crushed bricks (Thelen and Howe, 1978, p. 23).

A parking lot was constructed at the University of Delaware in Newark later in 1973 (Bachtle, 1974; Thelen and Howe, 1978, p. 23-24), designed by John Middleton and Edward R. Bachtle. The site was on a 2 percent slope. The soil infiltration rate was believed to be 0.6 to 1.3 cm/h. At the lowest corner an "L"-shaped trench was excavated to reach a more permeable soil layer, 4.6 m deep and 1 m wide, filled with 2-cm stone. The pavement was layered as shown in Figure 2.3. The surface course was 6.5 cm thick, with 5.5 percent asphalt. The base was 30 cm thick to hold water while infiltrating. The base was layered with small stones to bind the large stone and to allow the asphalt-laying machine to lay the paving on a smooth and maneuverable surface. The cost of installation was considered 20 percent less than that of an impermeable pavement due to the reduction in curbing and in storm drainage conveyances. Four years after installation the condition was reported as good, with closed pores in a 2-m^2 area where all cars applied brakes, and with superficial loosening of the top layer of aggregate in 15-cm patches where wheels of stationary cars had been turned. In the two decades since the parking lot was constructed, no repairs have been necessary (Gary Smith, University of Delaware, personal communication, September 3, 1993). Like all other lots on the campus, it has been vacuumed

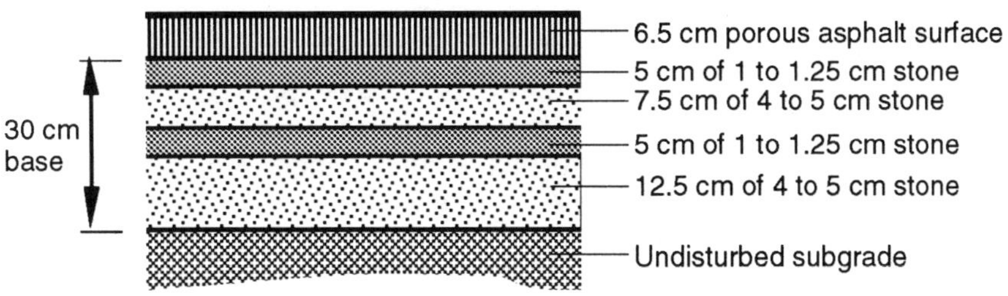

Figure 2.3 Porous asphalt pavement constructed in 1973 at the University of Delaware (after Bachtle, 1974).

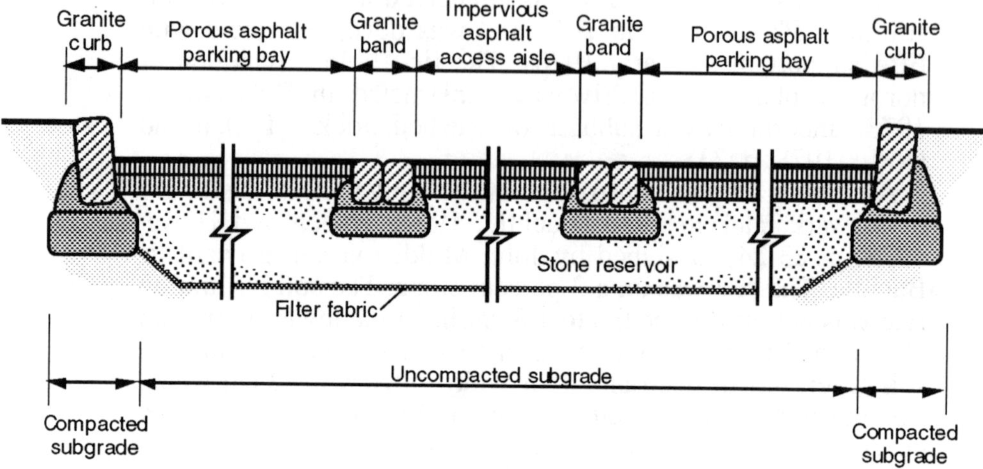

Figure 2.4 Porous asphalt parking lot constructed at the Morris Arboretum in Philadelphia (after Strom and Nathan, 1993, Figure 5.5).

about three times per year. During the winter, snow is plowed from the lot as necessary. To avoid clogging the pores, the maintenance crews, who refer to the lot as the "recharge parking lot," are careful to use only a light coating of rock salt and no sand.

The University of Delaware parking lot became the prototype for porous asphalt pavements that followed. Since 1973 hundreds or thousands of installations have been built in the United States, with the greatest concentration in the mid-Atlantic area. A frequently cited recent example is the parking lot at the Morris Arboretum in Philadelphia shown in Figure 2.4, designed by Andropogon Associates (Sorvig, 1993; Strom and Nathan, 1993, p. 109-203).

The permeability of porous asphalt concrete is high (Thelen and Howe, 1978, p. 13). At The Woodlands, Texas, under a 15-cm head, water passed into two porous asphalt pavements at rates of 53 and 84 cm/h. With high void space of 15 to 22 percent, infiltration rate can be 100 to 150 cm/h, and sometimes several times greater. The permeability of samples at the Franklin Institute was too high to measure reliably (>400 cm/h).

In field projects monitored during the 1970s, pores in porous asphalt remained open over all or almost all of the pavement surface (Thelen and Howe, 1978, p. 9-10 and 15). Loss of porosity was found in small areas due to three causes: 1) waterborne sediment draining onto the pavement and being ground in by traffic before being washed off; 2) soil dropped off trucks or tracked in on tires and subsequently ground in by traffic; and 3) sheer stress from many vehicles braking at the same spot, collapsing the pores. Compared to other asphalt concretes, porous asphalt is more subject to scuffing, such as occurs when the front wheels of stationary cars with power steering are turned. This effect is limited to the specific locations on the pavement where this kind of driving occurs; the area of pore collapse seldom exceeds 1 m^2 under each wheel. Small clogged areas do not necessarily harm the overall performance of a large pavement.

Methods of cleaning soil from the surface have been developed involving vacuum sweepers and high-pressure hosing (Nichols, 1992). At The Woodlands (Thelen and Howe, 1978), a pavement with an initial permeability of 53 cm/h was experimentally clogged to a permeability of 15 cm/h, then restored to full function by brushing, vacuum cleaning and high-pressure water washing. When a similar pavement was rendered completely impermeable with soil (to a depth of 1.25 cm), it could not be restored by cleaning. The only known method of restoring a completely clogged surface is by drilling small holes, for instance holes 0.5 cm in diameter spaced 30 cm apart.

Porous asphalt has proven to be a very satisfactory driving surface and component of urban communities (Thelen and Howe, 1978, p. 13-14). When wet it is more skid resistant than other pavements due to its coarse surface. It does not require curbs and gutters for primary drainage control. Puddles of measurable depth are essentially nonexistent. On a rainy night, markings painted on porous pavement are more visible than those on other pavements, because they are not covered by films of water. Noise levels of vehicles are slightly lower on a porous pavement than on others, perhaps because of the dissipation of tire sound energy in the void spaces.

Bachtle (1974) reported that, during severe ice storms and freezing rains, ice and snow conditions on the University of Delaware parking lot appeared the same as those on other pavements in the area. The porous asphalt froze over as did other pavement surfaces, with water running off the surface when rain fell in subfreezing temperatures.

Snow and ice may clear from porous pavement sooner than from impermeable material (Thelen and Howe, 1978, p. 20). When sunshine hits a dark asphalt surface, even through a few centimeters of ice, the pavement warms and melts the ice or snow from below. Meltwater drains through a porous pavement, eliminating the possibility of refreezing to form ice. The demand for deicing salts may be reduced.

Porous pavement is not damaged by freezing and thawing as long as the soil under it is non-heaving and drained (Thelen and Howe, 1978, p. 19). Existing porous asphalt installations have withstood 20 winters without heaving (Sorvig, 1993). Rain that falls on a very cold surface can freeze on the surface and in the near-surface pores. The pore space allows for expansion of some water without damage. In the base course, ice can develop only if water entered from rain or melting snow and then the weather turned very cold. Laboratory and field evidence is that the pavement is not damaged under such conditions due to the insulating properties of the porous asphalt, conduction from the relatively warm soil underlying the base course, the limited depth to which the frost line penetrates, the infiltration of some water before the base cools, the high latent heat of water making it resist freezing in near-freezing temperatures, and the absorption of some expansion by the stone voids.

A small amount of evidence suggests that aerobic bacteria can live in the soil under porous pavement (Thelen and Howe, 1978, p. 11-12 and 17). Aerobic bacteria can digest organic contaminants. How complete a purification will be done depends on the soil type, the supply of air and water, the temperature, the nature of the pollutant and the length of the percolate path. In the Woodlands experiment, an aerobic transformation of nitrogen forms was found. In the pavement's percolate, total organic carbon and chemical oxygen demand were low, apparently because of bacterial activity.

Porous concrete

Porous portland cement concrete pavement specifically for the infiltration of water was developed in the 1970s, primarily by Jack Paine of Florida. Previously, in the mid-1960s, an experimental road with a porous concrete surface layer had been constructed in England, and coarse-graded concrete had been used in lightweight building construction since the 1920s. Porous concrete's application in meeting Florida's stringent runoff control requirements has been monitored by the Florida Concrete and Products Association (no date). By about 1990, more than 90,000 m^2 of the pavement had been constructed in Florida, mostly in parking lots and lightly traveled roads. A public example is the parking lot of the St. John's Water Management District headquarters in Palatka (Sorvig, 1993).

Porous concrete (properly called "porous portland cement concrete") is constructed, like other concretes, of aggregate and a binder. The cementitious materials are the same as those in other portland cement concretes (Florida Concrete and Products Association, no date). Porosity is provided by the omission of fine aggregate. To provide a relatively smooth riding surface, and for adequate handling and placing properties, a maximum coarse aggregate size of 1 cm (no. 8 or no. 89) is often used. Pavement using this aggregate typically has a void content of 15 to 25 percent (Florida Concrete and Products Association, no date). The density of a pervious pavement is generally about 70 to 80 percent of that of other portland cement concretes, depending on aggregate source and degree of compaction.

The cement content in the water-cement paste is generally higher than that in other paving concrete. Water-cement content is controlled by testing on a unit weight basis. Too much water causes the paste to drain into the lower layers of the pavement, sealing the pores at the bottom and failing to bond at the top. Too little water results in incomplete hydration and poor binding of aggregate; in extreme instances this leaves the pavement with an agglomeration of loose particles.

The mixture sets in a short time, so rapid placement and compaction are essential. After the subgrade is compacted and just before placing the concrete, the subgrade is moistened, because porous concrete has minimal free moisture. The poured concrete is compacted to the correct elevation with a roller spanning between forms. Usually no additional finishing is attempted. The surface and edges are covered to retard evaporation, and kept covered five to seven days.

Porous concrete pavement, like other portland cement concretes, acts as a rigid slab, distributing structural loads over a large area and limiting the structural need for or required thickness of an aggregate base course. In Florida, the thickness of the concrete surface course for light traffic loadings is 12.5 cm;

heavier loadings require a thicker course. An aggregate base course can be added to increase total pavement thickness or to increase hydraulic storage. In practice the depth of the base has typically been determined by hydraulic storage requirement, not by mechanical load (Florida Concrete and Products Association, no date; Sorvig, 1993).

For a rigid slab, uniformity of subgrade support is more important than the subgrade's absolute strength (Florida Concrete and Products Association, no date). Due to the uniform infiltration of rain water under porous pavement, the subgrade is uniformly moist during both dry and wet subgrade conditions. This may explain the absence of pumping failures in pervious pavements constructed to date. Variable support may be experienced where the subgrade is highly compressible, lacks cohesion or expands when wet. Suspect subgrade materials must be individually analyzed for their support values and be modified, replaced or supplemented with additional depth of base course, as they would be for any concrete slab. Where pervious pavements adjoin impervious ones, the edges of the impervious ones have sometimes deflected and cracked due to nonuniform moisture at the edge. Alternations between the two types of pavement should be avoided in site layouts. Where such interfaces occur, the edges of the impervious pavements should be increased in thickness and reinforced.

The drying shrinkage of a porous cementitious mixture is substantially less than that of other concrete. This explains why some large installations using no control joints showed no visible shrinkage cracking after years of service (Florida Concrete and Products Association, no date). Apparently reinforcing to combat cracking is not necessary in porous concrete pavements in Florida. However, a conservative design would still include control joints spaced 18 m apart (Florida Concrete and Products Association, no date).

Permeability through one porous concrete surface was measured as 140 cm/h (Florida Concrete and Products Association, no date). Where improper mix or excessive vibration had reduced the porosity of the pavement surface, permeability was 30 to 40 cm/h (Wingerter and Paine, 1989, p. 5). An attempt to clog the surface with sand showed that permeability could be immediately restored with brooming; even when the surface was clogged by pressure washing soil into the pores, the pavement retained some permeability (Florida Concrete and Products Association, no date).

Where fine-textured subgrade limits infiltration rate, additional depth of base course can store water during infiltration. Permeability of the subgrade can be measured after compaction, using a double-ring infiltrometer or other technique, to confirm the needed depth of base. A base reservoir can be of open-graded (no. 57) stone, open-graded portland cement concrete, or sand having adequate void space. A base course can have lower cement content than the surface course (Florida Concrete and Products Association, no date).

Pervious pavement provides a smooth riding surface. Its friction value is higher than that of other pavements (Florida Concrete and Products Association, no date).

A survey of installations in Florida (Wingerter and Paine, 1989) found that subgrade conditions such as compaction and permeability did not change significantly with time after installation, even after many years and many wetting cycles. Occurrences of clogging of the pavement surface were insignificant.

Open-celled pavers

Open-celled pavers are precast or cast-in-place pavers made of concrete or plastic and containing open cells. The cells may be filled with soil sown with grass or filled with porous aggregate. The solid portion is intended to transmit structural loads, protecting the adjacent cells from traffic compaction and abrasion. Among the proprietary products that could be included in this category are meshes of plastic rings and wire turf reinforcement (Nichols, 1993c).

Open-celled pavers were first developed in Stuttgart, Germany in 1961 to reduce the "heat island" effect of large parking areas (Southerland, 1984). The idea was soon being used in parking areas, road shoulders, pedestrian ways and slope stabilization sites in many parts of Europe. They were introduced to the United States in the late 1960s and early 1970s, where they reinforced the earlier work of the U.S. Soil Conservation Service in applying cellular concrete blocks to stream bank erosion control. There are now a number of manufacturers located or represented in the United States.

Figure 2.5 Two types of open-celled pavers (after Southerland, 1984). On the left, lattice; on the right, castellated.

The products vary in size, weight, surface characteristics, strength, durability, interlocking capabilities, proportion of open area per grid, and runoff characteristics. Figure 2.5 shows two general types, lattice and castellated. Lattice pavers have a flat upper concrete surface through which the open cells penetrate. Castellated pavers have a complex surface of wafflelike hubs.

Lattice pavers provide large continuous solid surfaces for walking and biking. However, Nichols (1993a, 1993c) found that the most continuous grass cover and best appearance were maintained by castellated pavers with Bermuda grass. The shallow rhizomes by which Bermuda grass spreads were able to penetrate around castellations, creating a uniform turf surface. Lattice pavers and plastic ring meshes were less successful, apparently because the isolation of individual cells prevented spreading of turf from one cell where it is established to nearby cells.

Installations of open-celled pavers have been most common in lightly traveled areas such as overflow parking areas and emergency access ways. They have also been used as linings to protect drainage swales from erosion and as shoulders and parking areas along golf cart paths.

Most producers of concrete pavers are members of the National Concrete Masonry Association, which has established standards for its members' products including strength, water absorption, minimum thickness of grid walls or webs, response to freeze-thaw cycles, and uniformity. No similar set of standards yet exists for plastic pavers (Nichols, 1993c).

Pavers are commonly installed on a compacted subgrade with a coarse aggregate base up to 30 cm thick depending on load and soil conditions. The base may be eliminated in light residential applications where the subgrade is sufficiently stable. The base is covered with a sand setting bed. Topsoil is placed within about 1 cm of the grid surface and dampened to allow settling; then additional soil can be added. New seedlings should be mulched, at least in lattice pavers, because the small soil volumes in the cells are vulnerable to drying out (Southerland, 1984).

Another method of planting is to cover the entire pavement, whether lattice or castellated, with a few centimeters of topsoil and to seed the entire surface. Figure 2.6 shows a lattice parking pavement in Winter Park, Florida planted this way with St. Augustine grass. Its appearance is much more like that of a lawn than a pavement, except at a few corners where the edges of the pavers can be glimpsed under the turf layer. The only compacted places on the turf are the wheel tracks of cars where they pulled daily into their parking places.

Figure 2.6 Parking lot in Winter Park, Florida with topsoil over lattice pavers.

The installation cost of open-celled pavers tends to be higher than that of porous asphalt paving. Transportation of pavers to installation sites is a significant part of their cost. However, like other porous pavements, their cost may be less than that of impermeable pavements when the resulting reductions in drainage inlets and pipes are taken into account (Nichols, 1993a).

Grassed pavers require the same maintenance as lawns: mowing, fertilizing and, in some cases, irrigation, aeration and reseeding. Mowers may experience roughness across some pavement surfaces. If grass must be reestablished in cells compacted from heavy use, removing the soil from each cell manually may be required (Nichols, 1993a).

Day (1978) found in laboratory tests that, in the rational formula, open-celled pavers have runoff coefficients of from 0.05 to 0.35. The value depended primarily on subsoil condition and secondarily on slope and configuration of the paver surface. Castellated grids, with relatively little concrete at the surface, reduced runoff the most. Clogging of cells with compacted soil has not been reported in the literature (Nichols, 1993a), but crusting of poorly vegetated soil has been observed in the field.

Other materials

Figure 2.7 shows a variety of constructions using at least partly permeable surfaces.

Solid unit pavers allow infiltration if they are spaced to expose permeable material. Hade (1987) used a portable rainfall simulator to evaluate infiltration and runoff of installed pavers in Indiana under low rainfall intensities. The pavers were interlocking solid units of compressed concrete, but belonged to the family of porous pavements because they had sand joints and were placed on sand beds and aggregate bases. The ratio of runoff volume to rainfall volume was only 0.13 to 0.51 with half-hour rainfall intensities of 0 to 7 mm/h, increasing to 0.66 to 0.76 with intensities of 15 to 30 mm/h. Thus a substantial portion of rainfall infiltrated at very low rainfall intensities. The exact amount of infiltration and runoff varied with pavement slope and perhaps type and intensity of traffic load. However, Borgwardt (1993), performing similar experiments on stone setts in Germany, found that penetration of mineral and organic particles into narrow sand joints and the underlying sand setting bed can severely limit permeability.

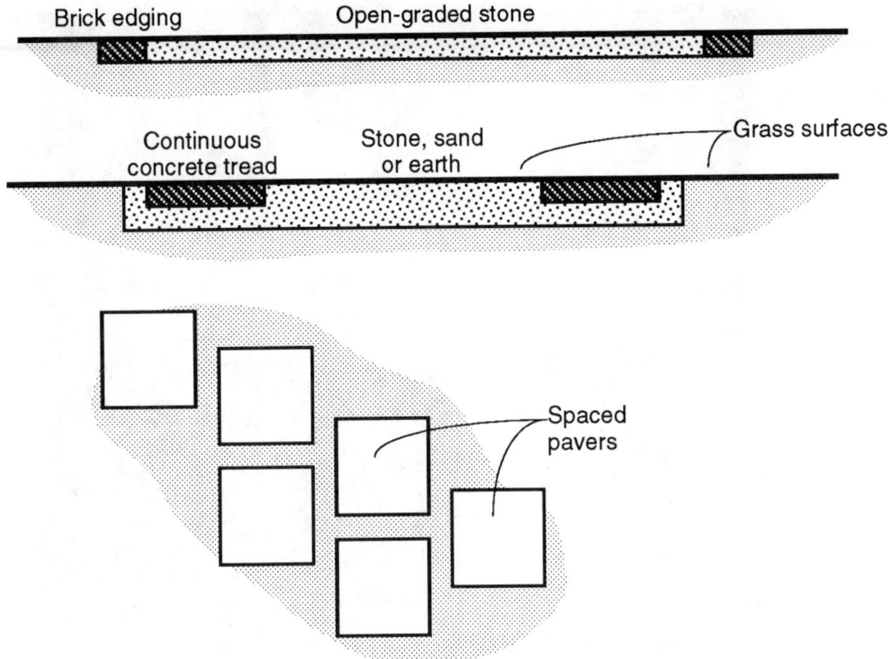

Figure 2.7 Constructions incorporating pervious surfaces. Top, a brick-lined gravel driveway; center, a driveway narrowed to two concrete bands; bottom, a walkway of spaced slabs.

Loose porous aggregates laid on the ground are permeable pavements for walking, jogging, biking, or very light vehicular traffic (Nichols, 1992). Examples are stone, shell, wood chips and bark. These materials can be relatively inexpensive to purchase and install. Figure 2.8 shows the simple gravel surface of a public plaza in downtown Orlando, Florida. On moderate slopes subjected to moderate traffic or concentrated runoff, loose materials can be displaced, and need to be raked back into place or replaced. Organic materials decompose over time and need to be replenished.

When aggregate of stone or pebbles is bound by an epoxy resin it is referred to as epoxy aggregate (Nichols, 1992). Installed cost can be high. It has been used as a permeable surface course over a base of porous asphalt, porous concrete or crushed stone. It is used mostly in pedestrian and light ve-

Figure 2.8 Level gravel pedestrian surface at Sunbank Plaza, Orlando, Florida.

hicular areas. It has been used in tree grates to allow infiltration and aeration of the root zone and in swimming pool decks to prevent surface ponding. The surface can be slippery, because the epoxy resin completely coats the particles; to add skid resistance a light dusting of clean white sand has been added to the surface before the epoxy hardens. This mixture can have higher compressive and tensile strength than most concrete pavements, depending on the aggregate used. It is resistant to deicing salts, petroleum compounds and weak acids.

Certain specialized playground, recreational and athletic surfacing products are porous (Nichols, 1993b). They include premanufactured rubberized mats and tiles and poured-in-place surfaces composed of rubber particles with a urethane binder. Most can be installed over a porous asphalt, concrete or aggregate base. The Safety Deck, made by Mat Factory Inc. of Costa Mesa, California, contains open cells for grass. Among the applications of such products are swimming pool decks, running tracks, and walkways around golf course clubhouses, where characteristics like brilliant colors and skid resistance are advantageous.

INFILTRATION BASINS

Where excess surface runoff is generated despite maximum feasible use of permeable surfaces, a basin can be constructed to capture and store the excess water while it infiltrates through the basin's floor and sides. Basins are built by excavation in the terrain, by placing dikes parallel to topographic contours, or by placing check dams in swales.

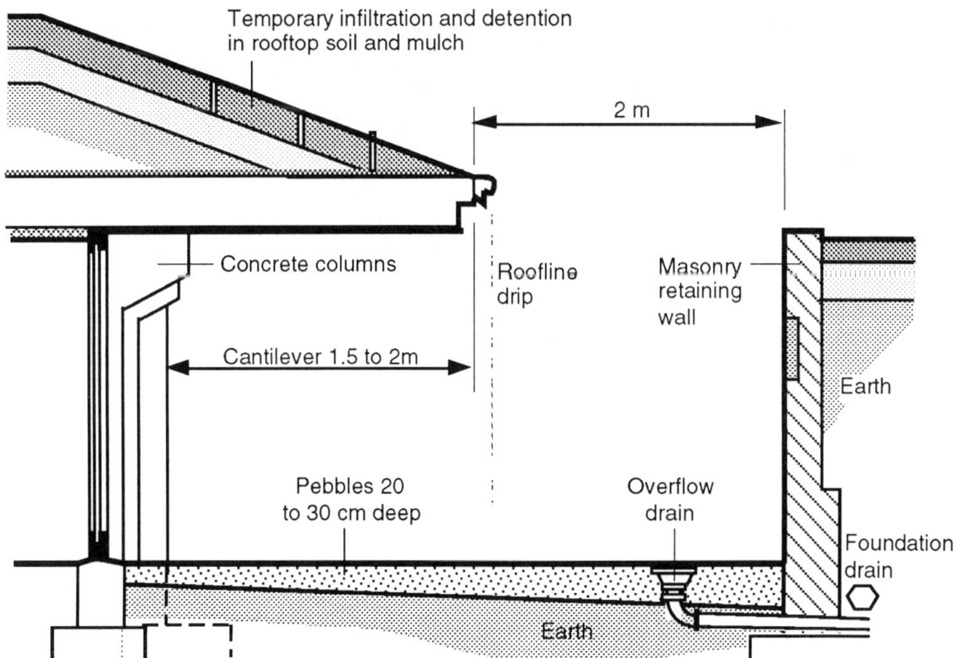

Figure 2.9 "Percolation bed" at the underground office of Malcolm Wells in Cherry Hill, New Jersey (after Wells, 1981, p. 27).

An example of a basin is on the site of Construction Fasteners, Inc. in Reading, Pennsylvania. Malcolm Wells (1971; 1974a; 1974b; 1981, p. 13-16) designed this site in 1966 and made successive additions over the years. The roofs and parking lots drain into a "percolation bed" surfaced with pebbles and interspersed boulders, looking like a sunken sculpture garden. Water occasionally ponds above the surface of the pebbles for a couple of days at a time. In 1974 the same architect installed the pebble courtyard shown in Figure 2.9 at his underground office in New Jersey. Despite the integration of the courtyard with the structure, no foundation problems have been reported.

It is possible for two or more small basins to do the hydraulic work of one large one. Small basins can fit inside the rights of way of streets and under parking lots, and can effectively utilize space within freeway interchanges and along highway shoulders. They can fit into the unused space under wooden decks and boardwalks. They can be placed at building downspouts to infiltrate roof runoff in the shadows of buildings.

If a prospective basin site has been artificially filled, saturating and destabilizing the base of a structural fill must be avoided. If the fill is not structural, the fill soil should be tested for hydraulic conductivity, or infiltration basins should be excavated below the base of the fill.

A basin's hydroperiod or long-term occurrence of water is established by hydrologic design as described in Chapter 4. A basin with a dry hydrologic regime has standing water only briefly following major storms and is dry the rest of the time. An ephemeral basin has a widely fluctuating water level, or only occasionally standing water. A wet basin has standing water at a more or less stable level, forming a permanent pond or wetland. Each type of regime has implications for water quality improvement, vegetation, and potential clogging mechanisms, as described in Chapters 5 and 6.

SURFACE BASINS

An open basin at the ground surface can be used where sufficient open space is available. A surface basin's construction involves little cost other than proper grading and planting. It is immediately and completely accessible for inspection and maintenance.

A surface basin's landscape design is greatly influenced by the expected occurrence of water, whether in a dry, wet or ephemeral regime (Ferguson, 1990 and 1991). Within any given basin, relatively different hydrologic regimes can be distinguished at different elevation zones. The highest zones are the least frequently flooded and have the most perennially dry soil. At the lowest elevation, runoff from every storm may be ponded at least temporarily, and sediment may be deposited relatively frequently.

Vegetation

Plantings in a surface basin can be native, ornamental or functional.

Different species have different types of adaptations to the hydrologic and substrate conditions created by different hydrologic regimes.

The importance of physical habitat to the health and survival of one kind of vegetation was evident in one basin on Long Island observed by Prill and Aronson (1978). A basin excavated in rapidly draining sandy soil had a lower floor flooded several times a month, and a bench level 0.3 m higher flooded only a few times a year. As a result of the difference in soil moisture, a dense cover of grass at the lower level contrasted with a sparse cover of grass at the bench level.

In basins with dry regimes, many recorded experiences have been with grass. For example, Hannon (1980, p. 136) reported that Bermuda grass (*Cynodon dactylon*) can withstand several days of submergence. In southern California, Bermuda grass has grown "luxuriantly" under prolonged soil wetting, provided the tops were not submerged (it died under prolonged total submergence), and survived long periods of drought (Muckel, 1959). Well-established Bermuda grows up through sediment deposits, forming a porous turf and preventing formation of an impeding surface crust. On Long Island, grasses in the fescue family such as creeping red fescue (*Festuca rubra*) and Kentucky 31 fescue (*Festuca arundinacea* or *Festuca elatior* var. *arundinacea*) have been commonly used because of their combination of hardiness, drought resistance and tolerance of brief flooding (Weaver, 1969, p. 48-49). The seeding mix includes a nurse grass such as perennial ryegrass (*Lolium perenne*) for early erosion control. Organic mulch is also added at the time of seeding, with an asphalt binder to keep it in place.

Table 2.1 Macrophytic vegetation surviving after 5 years of seasonal ponding in the Leaky Acres basin in Fresno, California (Nightingale and Bianchi, 1977).

SCIENTIFIC NAME	COMMON NAME
Polygonum coccineum	Swamp smartweed
Typha latifolia	Cattail
Sagittaria latifolia	Arrowhead
Ludwigia peploides	California water primrose
Lemna minor	Duckweed
Tillaia aquatica	Pigmyweed
Eleocharis macrostachya	Creeping spikerush

Mixtures of wildflowers were successful in 10 stormwater basins in New Jersey where ponding time was usually less than 24 hours (Oertel, 1993). The seedbeds were prepared by spreading 15 cm of topsoil and fertilizing according to soil tests. Several mixtures of annuals and perennials were seeded with noncompetitive nurse grasses. Seasonal overseeding increased diversity and continuity of flowering. Fall mowing eliminated woody growth. Establishment costs were higher for wildflowers than for turf but maintenance costs were lower.

In ephemeral basins, large seasonal water level fluctuation may severely limit the choice of vegetation. At the large Leaky Acres basin in Fresno, California, only the seven species of macrophytes listed in Table 2.1 lived in the wetted area after 5 years of operation characterized by ponding for months at a time in the summer with gradual periods of spring filling and fall draining (Nightingale and Bianchi, 1977). In Lubbock, Texas, an excavated infiltration basin with seasonal standing water and no special treatment for vegetation establishment had less diverse vegetation than undisturbed plots nearby; some of the major species present were *Kochia scoparia*, *Salsola kali*, *Solanum elaeagnifolium*, *Ambrosia psilostachya*, *Sorghum halepense*, and *Sporobolus cryptandrus* (van Hylckama, 1979). Similarly low diversity was observed in the Netherlands, in excavated basins where eutrophic Rhine water was occasionally recharged, and no special vegetative treatment was applied; major species present included *Urtica dioica*, *Cirsium arvense* and *Calamagrostis epigeios*.

A wet basin can be planted with hydrophytes, promoting wetland values such as wildlife. Contouring can create appropriate depth ranges for the desired submerged, emergent and floating-leaved plants. Obligate hydrophytes require water and cannot survive in dry areas; facultative hydrophytes may grow in upland areas as well as wet habitats. Large areas of tall emergent vegetation such as *Typha* or *Phragmites* may lead to anoxic conditions at the end of the growing season, when dying stems collapse into the water and their decomposition demands oxygen from the water. In Fresno, a massive growth of cattails and creeping rhizomes was removed by dredging; the consequent increase in water depth from 20 to 60 cm largely eliminated cattail regrowth (Nightingale and Bianchi, 1977).

Table 2.2 lists 80 plant species adapted both to irregular flooding (saturation or inundation to any depth, 5 to 12.5 percent of the growing season) and to drought—a combination of conditions likely to occur in many infiltration basins fed only by irregularly occurring stormwater. Garbisch (1992) listed these species as available for or successful in wetland planting projects, as defined by: 1) reports that they had been successfully planted in wetland environments (species colonized or developed from seed banks are not included), 2) listings in inventories of nurseries that produce wetland plant material, or 3) recommendations in wetland planting guides issued by agencies such as the U.S. Department of Agriculture. Additional species may be candidates for use, but inadequate published information prevented them from being listed. Most plants on the list are native to specific regions of the United States. Suitability of a species for a given basin or desired function may be influenced by pH and other soil characteristics, shade tolerance, rate and form of spreading, clumping and other growth habits, nutrient absorption, tolerance of seedlings to standing water, growth rate, nitrogen fixation, and value for specific faunal species. Loosestrife and cattail are considered invasive pest species in some areas.

In addition to the plants listed in Table 2.2, reed canarygrass (*Phalaris arundinacea*) has shown a remarkable combination of tolerance to inundation and tolerance to drought (Rice and Pinkerton, 1993). On the banks of streams and reservoirs it has survived flooding for weeks at a time, and seasonally low moisture when rainfall and water levels decline. It spreads by producing rhizomes, forming a dense sod.

Table 2.2 Plant species that have been successfully used in or are commercially available for wetland plantings (Garbisch, 1992, Table H-6).

SCIENTIFIC NAME	COMMON NAME
Herbaceous vegetation:	
Agrostis alba	Redtop
Andropogon glomeratus	Lowland broom sedge, bushy beardgrass
Andropogon virginicus	Broom sedge
Arisaema triphyllum	Small jack-in-the-pulpit
Asclepias incamata	Swamp milkweed
Aster novae-angliae	New England aster
Carex aperta	Columbia sedge
Carex lanuginosa	Wooly sedge
Carex lenticularis	Kellogg sedge
Carex retrorsa	Retrorse sedge
Carex stipata	Awl-fruited sedge
Carex trichocarpa	Slough sedge
Cyperus esculentus	Chufa, ground almond, yellow nutgrass
Juncus ensifolius	Swordleaf rush, three-stamened rush
Juncus tenuis	Slender rush
Juncus torreyi	Torrey rush
Leersia oryzoides	Rice cutgrass
Lythrum salicaria	Purple loosestrife, spiked lythrum, salicaire
Onoclea sensibilis	Sensitive fern
Osmunda cinnamomea	Cinnamon fern, buckhorn fern, fiddlehead
Osmunda regalis	Royal fern, flowering fern
Panicum virgatum	Switchgrass
Scirpus cyperinus	Wool grass
Thelypteris noveboracensis	New York fern
Thelypteris palustris	Marsh fern, meadow fern, snuffbox fern
Typha angustifolia	Narrow-leaved cattail
Typha latifolia	Broad-leaved cattail
Woodwardia areolata	Netted chain fern
Shrubs:	
Amorpha fruticosa	False indigo bush, indigo bush
Aronia arbutifolia (*Pyrus arbutifolia*)	Red chokeberry
Baccharis glutinosa	Mule fat, water-wally, seep-willow
Celtis occidentalis	Hackberry, sugarberry
Cephalanthus occidentalis	Buttonbush
Forestiera acuminata	Swamp-privet
Hibiscus moscheutos	Marsh hibiscus
Ilex decidua	Possumhaw, deciduous holly
Ilex verticillata	Winterberry
Magnolia virginiana	Sweetbay magnolia
Myrica pennsylvanica	Bayberry
Rosa californica	Wild rose
Rosa palustris	Swamp rose
Rubus spectabilis	Salmonberry

Table 2.2 (continued) Plant species that have been successfully used in or are commercially available for wetland plantings (Garbisch, 1992, Table H-6).

SCIENTIFIC NAME	COMMON NAME
Shrubs (continued):	
Rubus ursinus (*Rubus vitifolius*)	California blackberry
Sambucus canadensis	Elderberry, American elder
Spiraea douglasii	Douglas' spirea
Trees:	
Acer floridanum	Florida maple
Acer negundo	Box elder, ash-leaved maple
Acer rubrum	Red maple
Acer saccharinum	Silver maple, white maple, soft maple, river maple
Alnus rubra (*Alnus oregona*)	Red alder, Oregon alder
Amelanchier canadensis	Serviceberry, shadbush
Betula nigra	River birch
Betula populifolia	Gray birch, white birch, fire birch, oldfield birch
Carya aquitica	Water hickory
Chamaecyparis thyoides	Atlantic white cedar, false cypress, swamp-cedar
Cornus amomum	Silky dogwood
Cornus foemina racemosa (*Cornus racemosa*)	Graystem dogwood
Cornus sericea	Red-osier dogwood
Fraxinus caroliniana	Carolina ash
Fraxinus latifolia	Oregon ash
Fraxinus nigra	Black ash
Fraxinus pennsylvanica	Green ash
Gleditsia aquatica	Water locust
Gordonia lasianthus	Loblolly bay
Liquidambar styraciflua	Sweetgum
Nyssa aquatica	Water tupelo
Nyssa sylvatica	Black gum, black tupelo, sour gum
Persea borbonia (*P. palustris*)	Redbay, swamp bay
Pinus rigida	Pitch pine
Pinus taeda	Loblolly pine
Platanus occidentalis	Sycamore
Quercus bicolor	Swamp white oak
Quercus lyrata	Overcup oak, swamp post oak, swamp white oak
Quercus palustris	Pin oak, Spanish oak
Quercus phellos	Willow oak
Salix nigra	Black willow
Salix lasiolepsis	Arroyo willow
Salix nigra	Black willow
Taxodium ascendens (*T. distichum* var. *nutans*)	Pond cypress
Taxodium distichum	Bald cypress
Ulmus americana	American elm, white elm

Figure 2.10 Basin receiving runoff from drainage swales in the Village Homes community in Davis, California (photograph by Roger D. Moore).

Multiple pursuits

The number, size, shape, construction and planting of basins should be suited to the slope, configuration, human use and environmental relationships of each site. Figure 2.10 shows a sand-floored basin in the Village Homes community in Davis, California, integrated closely with a children's playground and nearby pedestrian ways.

Water's aesthetic roles can be symbolic, psychological or visual (Litton and others, 1974). Landscape character results from the combination of water, landform, vegetation and structures. In the overall arrangement of a site, a water feature can act as frame, spine, focus or open space. Cues to contouring, planting and materials can be taken from the surrounding landscape in order to blend a water feature into its community. Fulfilling the aesthetic intentions of a water feature may require adding plantings, structures, pedestrian pavements, lighting or semi-submerged rocks. If inflows of debris will be significant, such as paper or tree litter washed directly off a pavement, refined designs should not be attempted. In successional, unmaintained landscapes, decomposition and soil development can absorb litter deposits.

In a dry basin, turf or other low-growing ground cover can form an open space extending the composition of any other nearby open spaces. Nonliving ground covers could include sand and open-graded river stone. For a dry basin to have active recreational use, it must have adequate length and width, turf cover, and freedom from hazardous hard structures. Heavy recreational use can compact soil and reduce infiltration rate.

A permanent pool can be formed by locating an infiltration basin where the water table is shallow and its level steady, and excavating to expose it. Stormwater is temporarily mounded on top of the water table until it infiltrates through the basin sides, pushing down and spreading out the underlying water table. Alternatively, a permanent pool can be formed by allowing large quantities of overflow in wet periods to assure at least enough water to maintain the pool most or all of the year. In the Puget Sound area, a unique climate has allowed the establishment of permanent or near-permanent pools and wetlands with little overflow. Small rainfalls replace water losses almost daily. The pools support native plants and perform preliminary treatment for water quality, while the steadily standing water infiltrates into the underlying soil.

In wetland-rich Florida, ephemeral infiltration basins are interpreted by the public as a natural part of the environment. In California, ephemeral basins are often referred to as "percolation ponds;" moderate ponding time is easily tolerated in the semiarid climate. In other regions the mud, mosquitoes and safety hazards perceived to be associated with temporary pools may not be immediately tolerated by urban residents.

Runoff can be delivered to a surface basin by a vegetated swale, an armored flume or drop structure, or a buried culvert discharging directly to the basin floor. Inflow structures can be integrated with the form of the surrounding ground. Materials such as concrete and masonry can be selected and shaped for consistency with other nearby structures and the intended tone of the landscape design.

Hazards to human safety can be eliminated if they are recognized during design. In basins holding still, ponded water, accidents have been reported only where steep sides, covered with slippery wet grass, sloped directly down into deep water (Ferguson, 1991; Ferguson and Debo, 1990, p. 233). Basins can be designed to make falling and trapping of persons less likely and to make it easier to reach them when trouble occurs. Basin characteristics that can be modified for safety include water depth, objects in the ponded area, shoreline gradient and ground cover, and access control. For visibility the edge can be marked by a change in ground material or slope or a line of vegetation, boulders, logs, bollards or rails. Infiltration basins never incorporate the most dangerous feature of stormwater basins, a vortex at the submerged mouth of a culvert operating under inlet control. A broad, open overflow weir is unlikely to hurt anyone, because the hazard of moving water, when present, is clearly visible, and there is no underwater enclosure.

A basin intended to function for wildlife habitat requires interspersed water and lowland vegetation, often combining trees, shrubs, grasses and aquatics. Extensive edges and shallows are needed for feeding. Diverse spits, inlets and islands are needed for nesting, loafing and cover. At the large Leaky Acres basin in Fresno, where no special features were added for habitat values, the aquatic and semiaquatic bird species listed in Table 2.3 were observed. Artificial summer ponding at this site may be important to birds in the California climate, where natural wetland water levels are low in the summer (Nightingale and Bianchi, 1977). Bird population declined by 80 percent after a massive growth of cattail was removed and weeds were removed from adjoining slopes, leaving little but open water and bare soil. Ducks, black-necked stilts and American avocets particularly declined.

Fish require a permanent pool with a variety of microhabitats. They need some open water with limited aquatic vegetation and with scalloped edges to define defensible territories. Various bottom materials, depths and levels of shading are needed to provide diverse food and cover. Mechanical aeration might be needed at some sites.

Many species of algae may occur in infiltration basins, of the green, blue-green, euglenoid, yellow-green and yellow brown types (Nightingale and Bianchi, 1977). Algal growth relies on nutrients in a basin's substrate and in the inflowing runoff. If

Table 2.3 Bird species observed at Leaky Acres, Fresno, California (Nightingale and Bianchi, 1977).

SCIENTIFIC NAME	COMMON NAME
Actitis macularia	Spotted sandpiper
Anas cyanoptera	Cinnamon teal
Anas platyrhynchos	Mallard
Anthus spinoletta	Water pipit
Ardea herodias	Great blue heron
Aythya americana	Redheaded duck
Charadrius vociferus	Killdeer
Ereunetes mauri	Western sandpiper
Himantopus mexicanus	Black-necked stilt
Larus delawarensis	Ring-billed gull
Leucophoyx thula	Snowy egret
Limnodromus scolopaceus	Long-billed dowitcher
Megaceryle alcyon	Belted kingfisher
Podiceps caspicus	Eared grebe
Podilymbus podiceps	Pied-billed grebe
Pulica americana	American coot
Recurvirostra americana	American avocet
Totanus malanoleucus	Greater yellowlegs

nutrients were present in the soil before infiltration was initiated, and not replenished from fertilizers or other sources in the drainage area, they can be gradually leached from the soil by the infiltrating runoff. In such cases, algal growth may be vigorous during the first few years of operation and then decline to a steady state.

There are hundreds of species of mosquitoes in the United States (Arden Lea, University of Georgia Department of Entomology, personal communication, July 2, 1993). The method of breeding that may occur in a given infiltration basin depends on which species are present and on the environmental resources in the basin. Some species lay eggs on the still open water of permanent pools; the eggs hatch in two or three days. Others lay eggs in soil and leaf litter exposed in an infiltration basin by a receding water line, where the eggs hatch when the area is flooded again. After hatching, the larvae of all species dive into the water for food and come to the surface for air. They do this where the water is quiet, particularly where emergent and floating aquatic vegetation protects the water surface from turbulence. Thus mosquito breeding can be prevented by sizing the basin for either a dry or a wet regime and, if wet, by suppressing growth of aquatic vegetation or making the water a meter or more deep throughout the basin, including at the edges. Where larvae hatch despite efforts to suppress breeding, they can be suppressed by fish that eat mosquito larvae, such as mosquitofish (*Gambusia*). To support fish in an infiltration basin, the lowest part of the basin can be reserved as a permanent pool; the fish that survive there could temporarily move out into the rest of the basin when the water rises.

Flying insects, predominantly midges (*Chironomus* sp.), became a problem near residences on one side of the Leaky Acres facility, especially where there was outdoor lighting (Nightingale and Bianchi, 1977). Light trapping yielded 42 species of insects. The area of the basin with greatest flying insect population had soil very high in organic matter as a result of its prior use as an irrigated pasture. The organic matter was food for insect larvae. Midge growth was controlled by drying the basin and disking the surface. With time, the soil organic matter oxidized and declined, and the problem of flying insects subsided. Mosquitofish were placed in the basin for additional control; their effect, however, was small compared with the reduction of organic matter.

SUBSURFACE BASINS

A subsurface basin uses the void space of buried stone, pipe or manufactured chambers to create storage volume. The land surface can be reclaimed for other uses. Parking has been the most common surface use, being compatible with the large level surface of the basin and carrying only light vehicular traffic. Underground structures involve greater construction expense than surface basins, so they tend to be used where land is unavailable for surface basins.

Subsurface basins eliminate the possibilities of mud, mosquitoes and safety hazards sometimes perceived to be associated with surface ephemeral regimes. The absence of light and possible lack of oxygen may affect microbiotic growth. Subsurface basins can infiltrate high proportions of annual inflows without evaporative losses.

Reservoir

Reservoir aggregate must be open-graded, that is, of a narrow size range, so that the voids between particles are not filled by smaller particles. Stormwater moves freely through and is stored in the continuous void spaces until it infiltrates the surrounding soil. The stone particles need not be large; all open-graded gradations have similar void space of 38 to 40 percent as a proportion of total volume. ASTM (American Society of Testing and Materials) gradations with 38 to 40 percent void space are numbers 1, 2, 24, 3, 357, 4, 467, 5 and 57. Many other organizations such as the National Stone Association and state highway agencies follow the same gradation numbering system. To support a pavement or soil over coarse reservoir stone, a layer must often be added of smaller stone, permeable to water and large enough not to penetrate into the voids of the reservoir.

Perforated or slotted pipes and manufactured chambers can supplement aggregate for storing water. Although pipes and chambers require purchase and installation expense, they generate hydraulic capacity quickly: the interior of a pipe or chamber is 100 percent void. Pipes can quickly distribute water along the entire length of a basin, collect coarse sediment before it enters aggregate fill, and provide monitoring and maintenance access. Pipes in infiltration basins should follow any local practices applying to drainage pipes in general, including minimum diameter for maintenance access, amount of backfill cover and, in cold regions, laying the invert below the frost line. An example of a manufactured chamber is the Infiltrator, made in the shape of an open-bottomed arch by Infiltrator Systems Inc. of Saybrook, Connecticut, based on the firm's prior experience with septic leaching fields. Another is the Flo-Well, manufactured in the shape of a vertical cylinder by O-Well Products of Hyannis, Massachusetts.

Paul Thiel Associates (1980) described the type of construction shown in Figure 2.11. In the bottom 0.5 m or more of the basin, reservoir stone a few centimeters in size is placed to form a bedding. Perforated pipe is placed on the bedding and covered a minimum of 0.3 m on the sides and top, or up to the pea gravel level. A minimum of 15 cm of the pea gravel is laid over the reservoir stone and covered with layers of geotextile, polyethylene sheeting or compacted dense-graded aggregate to

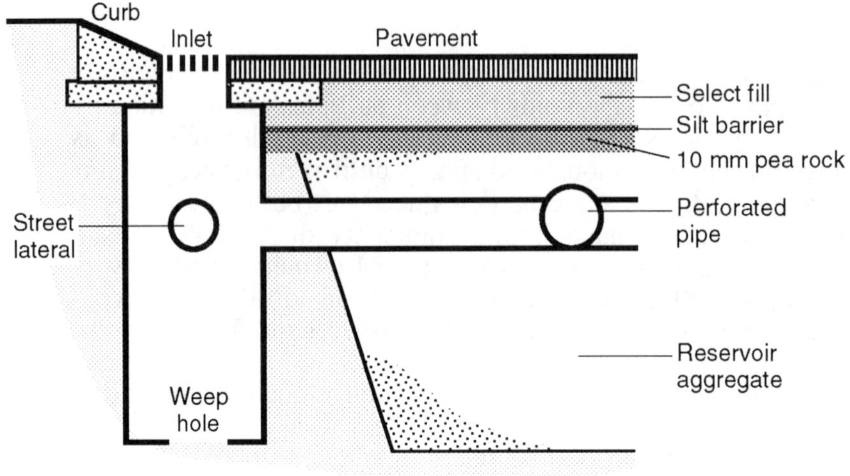

Figure 2.11 Infiltration basin under a pavement in Ontario (after Paul Thiel Associates, 1980, Figure 5-4).

bar movement of soil into the reservoir stone. Earth or base course is added above the barrier for even bearing of the pavement.

At Fairland Regional Park in Maryland, a coalition of public agencies used Infiltrator chambers in the basin shown in Figure 2.12 (Berg and Clement, 1992). The basin is located in a swale and receives concentrated flow through a stone surface layer along a 3.75 percent surface slope. A layer of chambers at the bottom of the basin increases the basin's void space, reducing the amount of stone needed.

In southern Florida, trenches are dug directly in pervious limestone and covered with reinforced concrete slabs. The walls are stable and permeable without additional structures or reservoir fill. A trench about 1 m wide can be dug with a backhoe or trencher. Wider trenches or basins are excavated with large equipment that enters the basin during excavation.

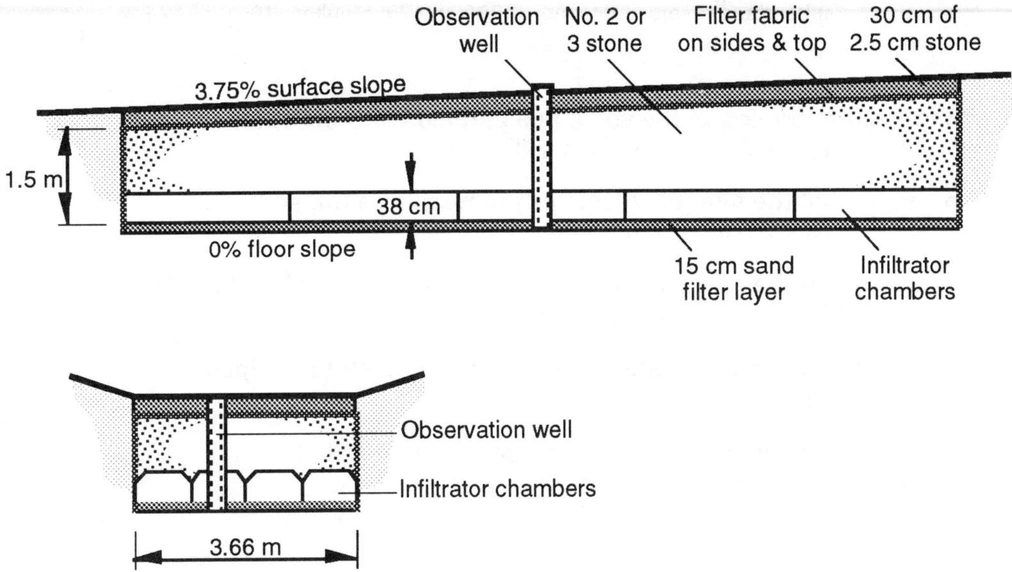

Figure 2.12 Basin with Infiltrator chambers at Fairland Regional Park in Maryland (after Berg and Clement, 1992).

Filter layers

Surrounding soil can move into an unprotected aggregate void space if the soil is finer than the aggregate and if certain types of flow conditions occur (Moulton, 1980, p. 98-101; Moulton, 1991, p. 12-52). Soil can move from the sides during recession flow (decline in water level following a storm) while the soil is still saturated: particles move into aggregate voids and are deposited on the floor, leaving gaps in the side walls. Soil can move from the roof (fill or aggregate over the open-graded reservoir) into the basin during recession flow, leaving gaps in the roof. Soil can move from the floor if the basin is excavated into ground water, if occasional artesian flow from below occurs, or if the aggregate at the floor is "pumped" under repetitive heavy traffic loads.

A layer of permeable sand or a filter cloth (geotextile) is sometimes placed between aggregate fill and the surrounding soil to prevent the soil from migrating into the aggregate voids

Table 2.4 Filter criteria to prevent soil from moving into the void space of a subsurface reservoir (Hannon, 1980, p. 170; Moulton, 1991, p. 12-52).

To prevent migration of the soil's fine particles into the voids:
D_{15} of filter ≤ 5 time D_{15} of the soil.

To ensure that the filter is at least as permeable as the soil:
D_{15} of filter ≥ 5 times D_{15} of the soil.

To ensure that the gradation curve of the filter is parallel to that of the soil:
D_{50} of filter ≤ 25 times D_{50} of the soil.

To prevent movement of filter aggregate through slots in pipes:
D_{85} of filter ≥ 1.2 times the slot width.

To prevent movement of aggregate through circular holes:
D_{85} of filter ≥ the hole diameter.

and reducing hydraulic capacity. Certain types of soils may need filters more than others, depending on cohesiveness and stability when wet.

A filter layer may be more necessary on the sides of an infiltration basin, where soil can collapse laterally, than on the bottom. Working of soil upward into a stone layer has been observed in some highways (Moulton, 1980, p. 34), but not under permeable pavements or at the base of infiltration basins. Permeable pavements and infiltration basins typically have base courses thicker than necessary for traffic loads alone, spreading out loads and reducing movement at the base, and tend not to be built where heavy traffic is expected. If the sides of a basin are gently sloping, as shown in Figure 2.13, perhaps a filter layer can be omitted altogether, and with it some construction expense and a potential uncertainty in basin conductivity.

Synthetic filter fabrics must be made of inert materials not susceptible to rot, mildew, attack by insects or rodents, or decomposition in ultraviolet light, alkalies or other environmental conditions. They must prevent the migration of fines without inhibiting the free flow of water (Paul Thiel Associates, 1980). Fabric installation must overlap at joints and avoid tearing or puncturing of the material.

Filter layers made of granular material must meet filter criteria for the specific soil conditions. Hannon (1980, p. 170) and Moulton (1991, p. 12-52) gave the examples of filter criteria listed in Table 2.4. The table uses a conventional notation for aggregate size, in which D_{15} is the particle diameter of which 15 percent of the material is finer, D_{50} is the diameter of which 50 percent is finer, and D_{85} is the diameter of which 85 percent is finer. To meet demanding soil conditions, sometimes more than one filter layer is needed, each layer meeting filter criteria for the next.

Surface layers

Surface materials over subsurface basins are selected to fit the surface use. Parking pavements are probably the most common type of surface. At the Sunbay Club, a residential part of Maitland Center near Orlando, a basin surface was constructed specifically to meet the demanding criteria of levelness, resilience and color for tennis. Figure 2.14 shows perforated pipes feeding trenches under the impervious tennis surface.

A basin can be installed most economically where the surface is literally level. Where a basin is constructed under a sloping surface, the depth of material at the upper end of the basin represents a partly wasted expense, because only the lowest part of the basin holds water. For example, in Figure 2.12, water that rises above 1.5 m overflows at the basin's lower end; the wedge of stone installed above the 1.5 m depth generates no functional hydraulic storage capacity. To accommodate a literally level surface, permeable pavement may be used, because it drains downward into the basin rather than requiring lateral

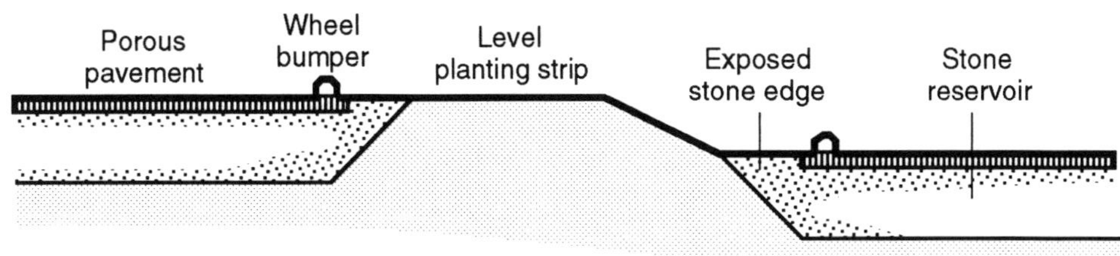

Figure 2.13 Stone edges and sloping sides of subsurface basins designed by Cahill and Associates and Andropogon Associates (after Sorvig, 1993).

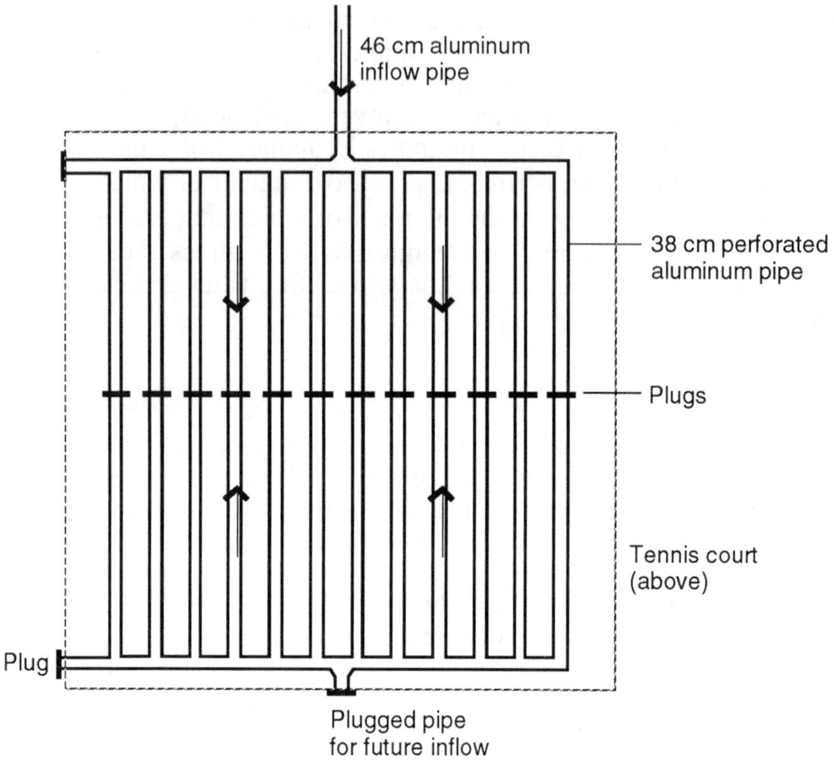

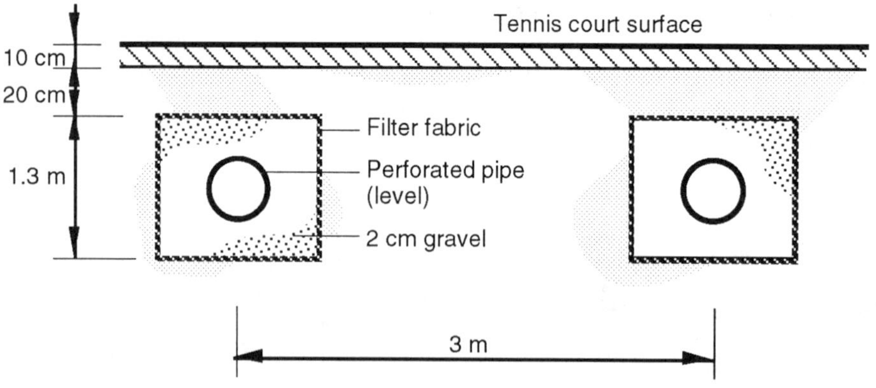

Figure 2.14 Infiltration under a tennis court at the Sunbay Club, near Orlando, Florida, designed by Post, Buckley, Schuh and Jernigan.

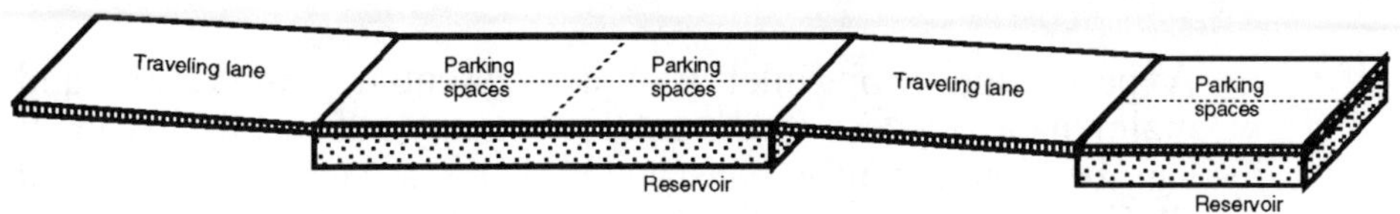

Figure 2.15 Formation of level areas for infiltration basins under a parking lot.

surface drainage. A literally level surface requires going outside old conventions that require positive slope on every urban surface. The old conventions were developed for impervious surfaces that drain only laterally, and did not take into account the possibility of vertical drainage or of runoff infiltration as a positive objective of grading design. To obtain level surfaces, basins must be laid out at the earliest stages of site planning to fit site topography and the grading demands of nearby drives and building entries.

Where a permeable pavement unavoidably slopes, such as along a long street or driveway, the water in the base course tends to flow to the low areas, leaving the upper parts of the reservoir empty of water. At low points and intermediate levels in the alignment, the profile can be made level, and effective infiltration basins can be installed under the pavement. Figure 2.15 illustrates the application of the principle to a large parking lot. The sloping parts of the pavement, where the base transmits water without storing it, need be no thicker than structurally necessary to handle traffic loads.

In unpaved areas, reservoir stone can be topped with a selected surface material. The stone surface in Figure 2.12 is appropriate for its location in a swale because it is nonerodible and highly permeable to inflowing runoff. In locations closer to human use, surface materials such as river stone can be selected for visual, symbolic or psychological roles. Pedestrian traffic can be borne with a simple boardwalk structure or preformed slabs laid directly on the stone fill. A basin covering of soil and turf was seen at an office site in Columbia, Maryland; the turf appeared slightly drier and thinner than surrounding turf, perhaps because of underdrainage by the reservoir.

Inlets

A perforated drop inlet provides essential observation and maintenance access to the interior of any subsurface basin. It can be conveniently constructed from a vertical perforated PVC pipe penetrating the full depth of the basin and capped with a circular metal grate. To avoid crushing of the PVC under a direct wheel load, the inlet can be located off vehicular pavements in a swale as shown in Figure 2.16.

If an inlet is located at the low point (sump) of a basin with a slightly sloping floor, the bottom of the inlet will be the first and last place that sediment and standing water occur. Pumping or otherwise removing sediment through the inlet would attack the deepest concentration of sediment in the basin. If the sump is a continuous trench where inlets are located at intervals, the basin can be conveniently flushed of sediment by pumping water in at one inlet and out at the next. Flushing of a large basin could be aided by a horizontal perforated pipe at one edge, distributing water across the entire basin floor for flushing down to a low collection point. The floor must be nonerodible during such a procedure; a stable filter layer or filter fabric may be required.

The same inlet can be used to set a basin's overflow elevation by placing the inlet's rim slightly lower than nearby pavements or other surfaces that are to be protected from overflowing water. The overflow would operate when a very large storm occurs or the basin is clogged (Moulton, 1980, p. 34). In the arrangement shown in Figure 2.16, the inlet and swale provide controlled overflow from the basin.

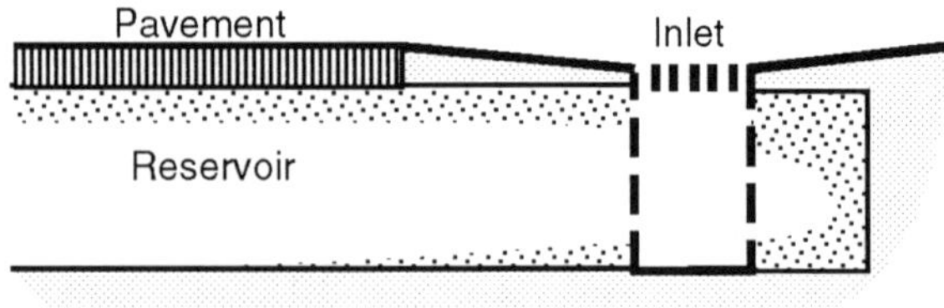

Figure 2.16 Inlet to a subsurface basin, located off the pavement in a swale.

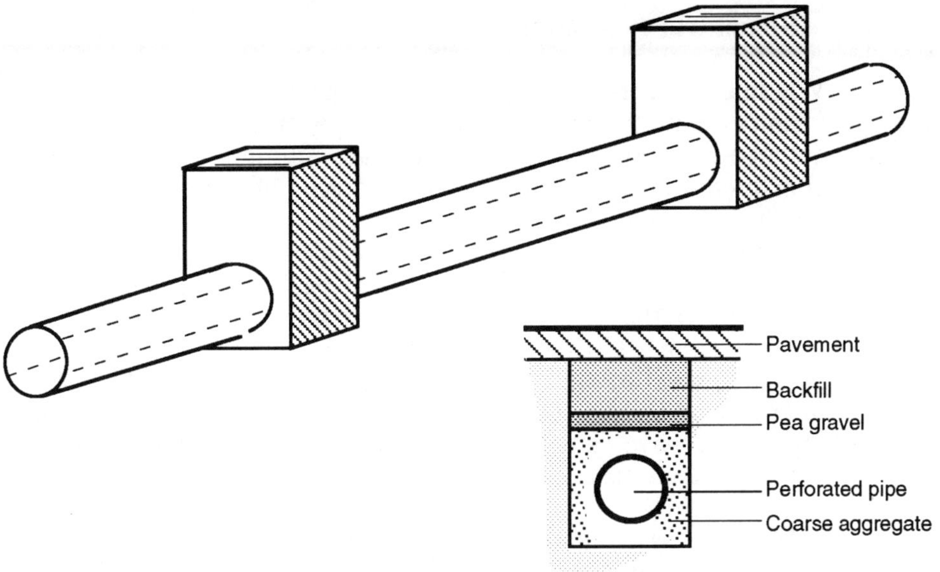

Figure 2.17 Inlets and perforated pipes feeding subsurface basins under city streets in Miami (after Hannon, 1980, p. 154).

For a basin under an impermeable surface, an inlet is a way to route runoff into the basin. Figure 2.17 shows a system of drop inlets used in Miami to penetrate impervious city streets and feed trench basins through perforated pipes.

For a basin under a permeable surface, an inlet provides a backup inflow route if the surface should become clogged. Cahill and Associates and Andropogon Associates construct their porous pavement installations with a stone trench around all edges as illustrated in Figure 2.13. This provides an alternative route for pavement runoff to enter the reservoir if the pavement is clogged, but it does not provide maintenance access to the soil surface as would an inlet.

A number of inlets feeding a single large basin, especially those joined by pipes, might maintain aerobic conditions in the basin substrate. This might affect microbial growth and the possibilities of water quality treatment and microbial clogging, the implications of which are described in Chapters 5 and 6.

Structural evaluation

Where a subsurface basin underlies a pavement carrying a mechanical load, the layers of basin and pavement must work together to support the anticipated load. Strength of a properly constructed layer is a function of the type of material and the depth of the layer.

Where porous reservoir aggregate is used as the base material in a pavement, a properly selected aggregate is essentially identical to the aggregate used in pavements designed for strength alone. Crushed stone must be specified and not uncrushed gravel because of the stability given by crushed stone's interlocking angular facets. The depth of the base course in an infiltrating pavement is likely to be greater than that in other pavements because of its dual function as a stormwater reservoir.

Moisture is a common occurrence in almost any kind of pavement. Soils under "impermeable" pavements become wet during wet weather due to cracks in the surface layer, percolation through the shoulders and artesian flow from below (Thelen and Howe, 1978, p. 11). That moisture is anticipated under any pavement is shown by the fact that subgrade strength, as defined by the California Bearing Ratio, is measured on saturated soil. In noninfiltrating pavements, drainage is provided by outletting the base to the surface. In pavements underlain by subsurface infiltration basins, primary drainage is vertical, through the soil.

A drainage blanket is a horizontal pavement layer specifically designed for removal of water in order to preserve the structural integrity of the overlying layers (Moulton, 1980, p. 87-98; Moulton, 1991, p. 12-29). A drainage blanket requires an adequate thickness of permeable material and a positive outlet for water. During design conditions the water surface must not rise above the top of a drainage blanket, although saturation of the layer is permissible for a few hours. Under a noninfiltrating pavement, the permeability, slope and rate of flow through a drainage blanket are analyzed to determine whether it meets this criterion.

The zone above an infiltration basin's overflow elevation can be considered a drainage blanket (Moulton, 1980, p. 30; Moulton, 1991, p. 12-35). The discharge from an infiltration basin drainage blanket is downward, into the void space of the reservoir, and also lateral, through the overflow outlet. Saturation of aggregate above a basin's reservoir can be expected only during the temporary peak of an extremely long and intense storm event, when the basin is full of standing water, the overflow is in operation, and water has risen in the overflow channel. Following subsidence of water in the overflow channel and decline of the water level in the basin due to infiltration, seepage out of the drainage layer into the basin is immediate due to direct vertical drainage under the full width of the drainage blanket.

Thus a properly constructed infiltration basin is not an addition of a soggy unstable base to an overlying pavement. It is instead a superlative underdrain. Infiltration basins can be coherent components of stable pavement construction.

Wells

The term "well" describes a subsurface basin that is relatively deep in proportion to its horizontal dimensions. An advantage of a well is that it operates at high head and therefore high infiltration rate near the bottom when the well is full.

In Maryland, small "dry wells" receiving runoff only from single roof downspouts have been successful in operation, perhaps because they receive very little sediment from their runoff source. Figure 2.18 shows a dry well at Maryland's Fairland Regional Park (Berg and Clement, 1992). The inlet pipe receives flow from a pavilion downspout; the building and down-

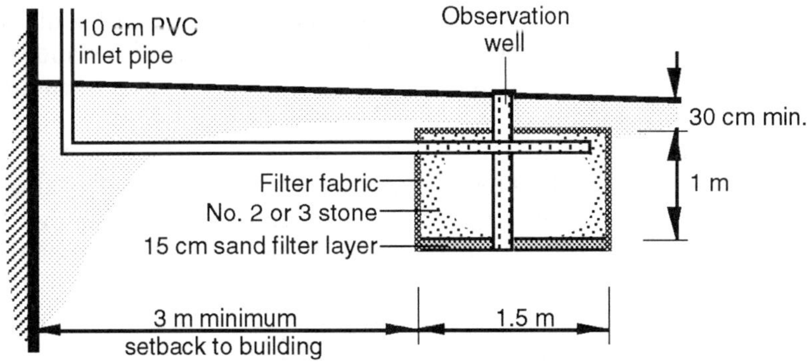

Figure 2.18 "Dry well" receiving roof runoff through a downspout at Fairland Regional Park in Maryland (after Berg and Clement, 1992).

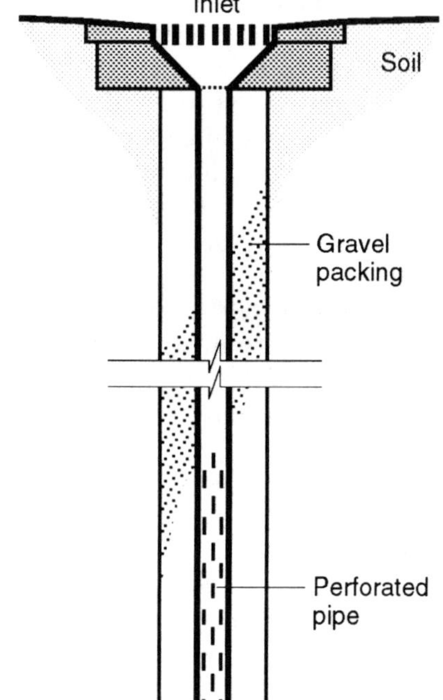

Figure 2.19 Narrow well (after Paul Thiel Associates, 1980, Figure 5-2)

spout were specified in separate architectural drawings. The basin's floor is 1.5m x 3m in area and level. The observation well sits on a metal foot plate.

Narrow wells like that in Figure 2.19 have been used for removal of standing water in isolated areas otherwise difficult to drain. Similar wells have been built in highway medians and on road shoulders for highway runoff. Simple wells have been augured with diameters of 0.5 m to 1.5 m.

Wells have been installed in the floors of surface and subsurface basins to supplement hydraulic capacity and infiltration rate. In California, wells like that in Figure 2.20, filled with gravel or sand, are used to penetrate shallow slowly permeable strata or to increase head or side surface area. A layer of filter aggregate is placed over the top of the well fill and is replaced if it becomes clogged (Jackura, 1980, p. 21).

Casings have been added to infiltration wells, making their construction similar to that of water supply wells. Cased wells have been hydraulically effective when inflowing water is clear of sediment and the installations are properly maintained. Cleaning or restoring wells clogged with sediment can be difficult. Construction and operating costs per unit volume of water drained are high compared with those of shallow surface basins (Jackura, 1980, p. 6).

In California, a casing usually consists of metal pipe (Jackura, 1980, p. 17-18 and 21). The pipe is perforated in zones of permeable soil strata where water is to be infiltrated. It is not perforated in other zones so as to minimize the amount of soil washing into the well. Two types of construction are shown in Figure 2.21. One is an open-end casing, in which compressed air is used to blast out a cavity of several cubic meters in the soil beneath the end of the shaft. This type has functioned well in the sandy soil of the Fresno area. A second type is the sand-packed perforated metal pipe, in which a smaller-diameter pipe is inserted into the casing and the space between them is filled with sand.

It is difficult to predict infiltration rates of deep wells accurately (Jackura, 1980, p. 21). Construction and testing of a prototype well would be more reliable than predictive modeling. If wells are installed in clusters, the effect of interference among nearby wells should be considered in estimating composite infiltration rates. Deep infiltration wells may be subject to regulation under environmental standards for subsurface injection of fluids.

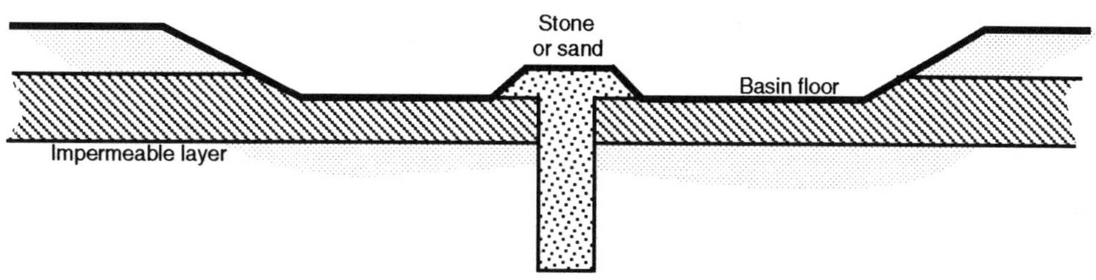

Figure 2.20 Well in the floor of a surface basin in California.

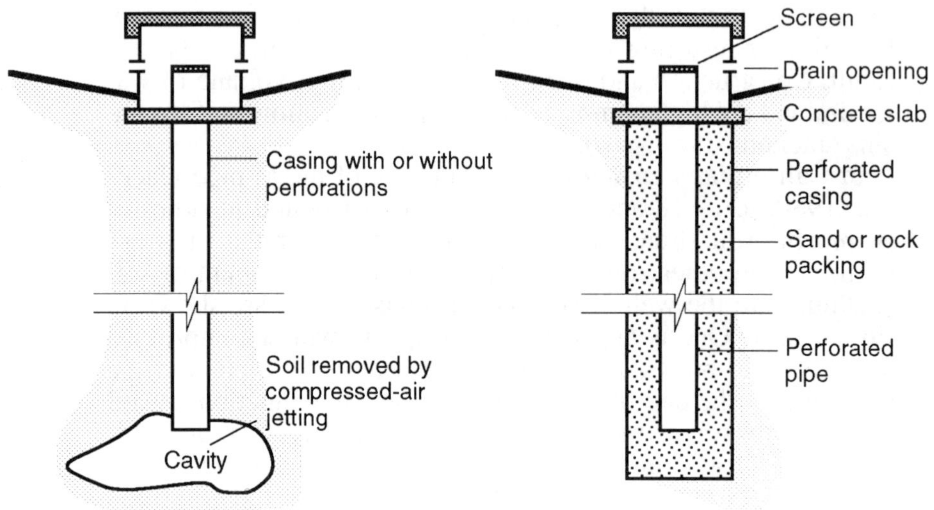

Figure 2.21 Two infiltration well constructions in California (after Hannon, 1980, p. 162). On the left, open-end casing; on the right, sand-packed perforated pipe.

On Long Island, cased wells like that in Figure 2.22 are installed in the floors of basins that have become flooded because of underlying strata of low permeability or clogged by accumulations of sediment (Aronson and Seaburn, 1974). They are used mostly in basins draining highways and industrial and commercial areas where no adjacent land is available for construction of another surface basin. They are commonly constructed of 3 m diameter precast reinforced concrete sections backfilled with coarse sand and gravel and installed at sufficient depth to penetrate the restricting layer or 2 m below the water table. Rectangular openings in the wall, amounting to 9 percent of the wall area, provide drainage. The shaft is given a concrete cover with a grate opening, and a graded filter is placed on top to prevent solids from entering. High cost and low exfiltration efficiency make such wells the least desirable of infiltration methods on Long Island (Hannon, p. 17-18).

CONSTRUCTION PROCESS

Vigilant control of construction and frequent postconstruction field inspection and refinement are essential for the long-term performance of infiltration surfaces and basins, as they are for any kind of environmental design.

The series of photos in Figure 2.23 illustrates the construction of a subsurface basin in Maryland. This basin is a small "dry well," intended to receive the runoff only from a single downspout. In A, the basin has been excavated and the soil surface into which water will infiltrate is visible; the shallow trench where a downspout will enter is visible on the left. In B, the floor and sides have been lined with a filter fabric, and a perforated PVC observation well has been set upon a metal floor plate. In C, the basin has been filled with open-graded aggregate, in this case a rounded pea gravel because no structural load will be imposed on top of the basin. The fabric will be lapped over the stone to finish the reservoir. In D is shown the construction detail that the basin is intended to execute, including a surface layer of soil.

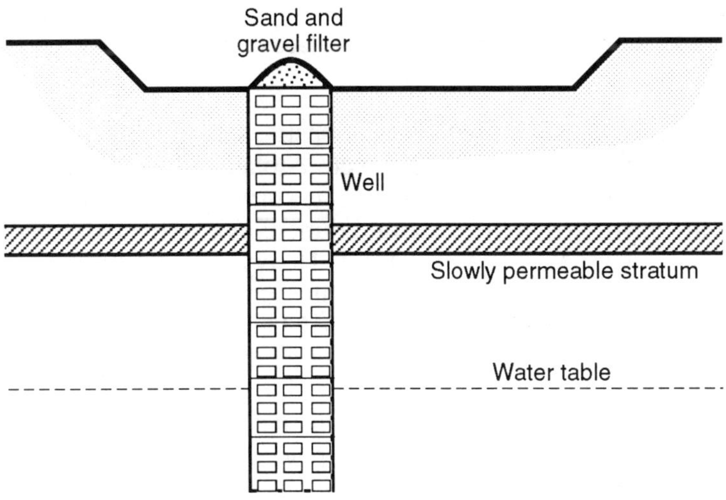

Figure 2.22 Cased well in the floor of a surface basin on Long Island (after Aronson and Seaburn, 1974).

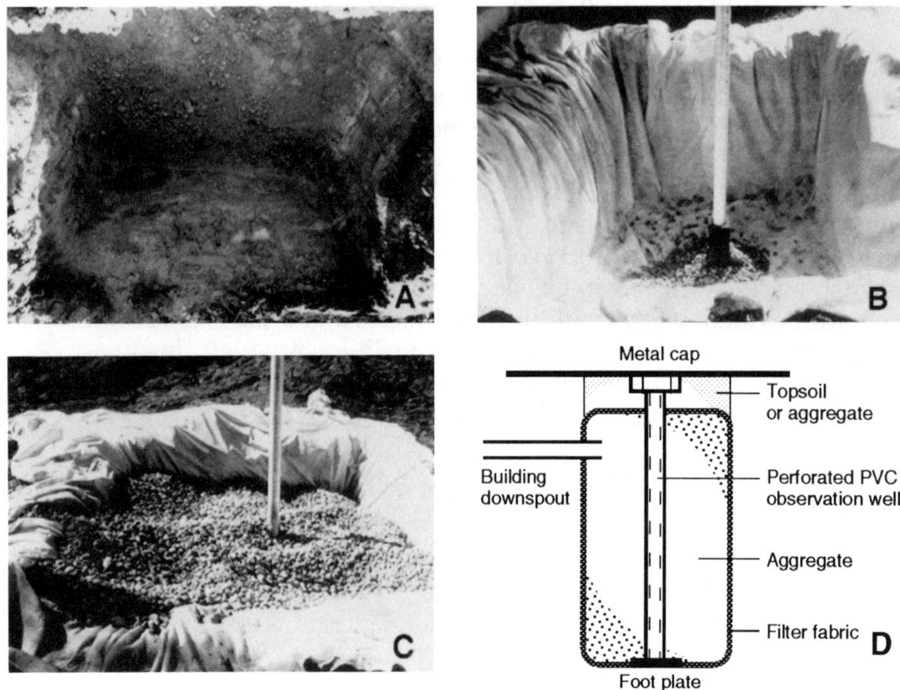

Figure 2.23 Construction of a small subsurface basin in Maryland. See text for details.

In Figure 2.24, a much larger basin is being built, but the same principles are being followed. A black filter fabric is visible on the edges of the stone. The surveyors are confirming that the basin meets the specified dimensions and elevations for hydraulic capacity and overflow. Heavy construction equipment has greatly disturbed the basin floor; this really ought to be prevented, but, as discussed in Chapter 6, this temporary disturbance may not be as important as the development of the soil surface after the basin is in operation.

Reservoir stone should be placed in lifts, with the delivery trucks driving only on previously placed stone. Vehicles must not track silt into the reservoir. If dirt is spilled or washed onto an infiltration basin, it must be vacuumed off where it appears.

Timing of construction of an infiltration basin within the site development of which it is part is of the utmost importance in preventing the basin from being damaged by inflowing sediment. Premature use of a finished basin to receive runoff, while surrounding slopes are unvegetated and erodible, would load the basin with sediment, reducing its hydraulic capacity. Although a basin should be excavated as early as necessary in project construction to obtain necessary fill, it is neither necessary nor desirable to complete it for runoff disposal until the project nears completion (Weaver, 1969, p. 47-50). Specifications must state the earliest point in progress when stormwater drainage may be freely directed to the basin and the means by which the opening to drainage is to be delayed.

A clear failure of an infiltration basin occurred in an Atlanta suburb where a developer was forced to turn from the locale's conventional detention to infiltration because he could not obtain a drainage easement for discharging onto downstream properties. A stone-filled infiltration basin was built at the lowest point of the site. It was surrounded by a dike made of the same open-graded stone as the reservoir. Residential construction continued upstream without any sediment control. The open voids of the dike failed to prevent sediment from reaching the basin and eradicating the basin's hydraulic capacity. The basin turned into an ephemeral surface pond with illegal discharges. By observing such failures, it is easy to learn their les-

Figure 2.24 Construction of a large subsurface basin in Maryland.

sons. Any combination of upstream at-source sediment control, proper silt fences around the basin, and staged basin construction would have preserved the basin's capacity and ensured that it would function as designed.

The most convenient way of accommodating infiltration basins on active construction sites is to use them temporarily as sediment traps for control of non-point source pollution. This application requires careful attention in implementation. In the rough grading stage of the overall project, the basins are excavated to about 0.3 m of their final floor elevations. The partially excavated basins can be used as sediment traps for the remainder of site construction.

Excavation to final dimensions must be deferred until the last possible step in the development, after buildings and pavements have been installed and all slopes in the drainage area have been at least seeded and stabilized with a reliable interim mulch or erosion control fabric. Sediment accumulated in the basin, if any, is removed with the final excavation. The excavated material is disposed of, and the installation of the basin's stone, structures and plantings begins.

From this time on the basin must be treated very sensitively to protect its capacity and conductivity. Compaction of the basin floor must be limited by using light construction equipment and minimizing the number of trips across the excavation. Inflow of residual sediment must be severely limited by constructing temporary diversion ditches on the upslope side of a basin or surrounding it with silt fences. As described in Chapter 6, an application of gypsum to the basin floor can improve soil structure, while mulch and vegetation can protect the floor of a surface basin from the disaggregating effects of falling raindrops. Only when the basin and its drainage area are stable can the basin be put into its permanent use, freely receiving clean stormwater from the watershed.

On Long Island, many basins have been constructed more or less along this procedure. They are allowed to receive large sediment loads during housing construction. After construction is completed, when vegetation is established and sediment inflows have ceased, the sediment accumulated in the basins is scarified, that is, broken up or loosened, and a thin surface layer is sometimes removed from the floor. Such basins are reported to operate well for many years afterwards (Aronson and Seaburn, 1974, p. 18-19).

REFERENCES

Aronson, D.A., and G.E. Seaburn, 1974, *Appraisal of Operating Efficiency of Recharge Basins on Long Island, New York, in 1969*, Water-Supply Paper 2001-D, Washington, D.C.: U.S. Geological Survey.

Bachtle, Edward R., 1974, The Rise of Porous Paving, *Landscape Architecture* vol. 65, Oct. 1974, pages 385-387.

Berg, Vincent H., and Paul F. Clement, 1992, Maryland's Stormwater Park Brings Control to Runoff, *Land and Water* vol. 36, Nov.-Dec. 1992, pages 26-31.

Borgwardt, Sonke, 1993, Versickerung auf Durchlassig Befistigten Oberflachen, *Wasser + Boden* 1993, no. 1, pages 38-41.

Bormann, F. Herbert, Diana Balmori, and Gordon T. Geballe, 1993, *Redesigning the American Lawn, A Search for Environmental Harmony*, New Haven: Yale University Press.

Day, Gary E., 1978, *Investigation of Concrete Grid Pavements*, Blacksburg: Virginia Polytechnic Institute and State University.

Diniz, Elvidio V., 1980, *Porous Pavement, Phase 1—Design and Operational Criteria*, EPA-600/2-80-135, Cincinnati: U.S. Environmental Protection Agency Municipal Environmental Research Laboratory.

Ferguson, Bruce K., 1990, Role of the Long-Term Water Balance in Design of Multiple-Purpose Stormwater Basins, pages 161-170 of *Proceedings, Council of Educators in Landscape Architecture Annual Conference*, Sept. 7-9, 1989, Amelia Island, Florida, Washington, D.C.: Landscape Architecture Foundation.

Ferguson, Bruce K., 1991, Taking Advantage of Stormwater Basins in Urban Landscapes, *Journal of Soil and Water Conservation* vol. 46, no. 2, pages 100-103.

Ferguson, Bruce K., and Thomas N. Debo, 1990, *On-Site Stormwater Management, Applications for Landscape and Engineering*, second edition, New York: Van Nostrand Reinhold.

Florida Concrete and Products Association, [no date], *Pervious Pavement Manual*, Orlando: Florida Concrete and Products Association.

Garbisch, Edgar, 1992, *Wetland Restoration, Enhancement, and Construction* [notes from American Society of Landscape Architects seminar], St. Michaels, Maryland: Environmental Concern Inc.

Girling, Michael J., 1992, Stadium Drainage Without the Water, *Stone Review* (National Stone Association) Dec. 1992, pages 32-33.

Hade, James D., 1987, Runoff Coefficients for Compressed Concrete Block Pavements, *Landscape Architecture* vol. 77, no. 3, pages 102-103.

Hamilton, George W., 1990, *Infiltration Rates on Experimental and Residential Lawns*, M.S. thesis, State College: Pennsylvania State University.

Hannon, Joseph B., 1980, *Underground Disposal of Stormwater Runoff, Design Guidelines Manual*, FHWA-TS-80-218, Washington, D.C.: U.S. Federal Highway Administration.

Jackura, Kenneth A., 1980, *Infiltration Drainage of Highway Surface Water*, FHWA/CA/TL-80/04, Washington, D.C.: Federal Highway Administration and Sacramento: California Department of Transportation.

Jones, D. Earl, 1967, Urban Hydrology—A Redirection, *Civil Engineering* August 1967, pages 58-62.

Litton, R. Burton, Robert J. Tetlow, Jens Sorensen and Russell A. Beatty, 1974, *Water and Landscape, an Aesthetic Overview of the Role of Water in the Landscape*, Port Washington, New York: Water Information Center, Inc.

Marvin, Robert E., and Associates, 1978, Jones Bridge Headquarters, *Landscape Architecture* vol. 68, no. 4, pages 294-295.

Marvin, Robert E., and James Paddock, 1979, A Corporate Headquarters Achieves Minimum Landscape Impact, *Landscape Architecture* vol. 69, no. 1, pages 70-73.

Moulton, Lyle K., 1980, *Highway Subdrainage Design*, Report No. FHWA-TS-80-224, Washington, D.C.: Federal Highway Administration.

Moulton, Lyle K., 1991, Aggregate for Drainage, Filtration, and Erosion Control, Chapter 12 of *The Aggregate Handbook*, Richard D. Barksdale, editor, Washington, D.C.: National Stone Association.

Muckel, Dean C., 1959, *Replenishment of Ground Water Supplies by Artificial Means*, Technical Bulletin 1193, Washington, D.C.: U.S. Department of Agriculture, Agricultural Research Service.

Newville, Jack, 1967, *New Engineering Concepts in Community Development*, Technical Bulletin 59, Washington, D.C.: Urban Land Institute.

Nichols, David B., 1992, Paving, pages 69-138 of *Materials,* Volume 4 of *Handbook of Landscape Architectural Construction*, Scott S. Weinberg and Gregg A. Coyle, editors, Washington, D.C.: Landscape Architecture Foundation.

Nichols, David B., 1993a, Exploring Alternatives to Porous Asphalt, *Landscape Design* vol. 6, no. 10, pages 8-15.

Nichols, David B., 1993b, Fresh Paving Ideas for Special Challenges, *Landscape Design* vol. 6, no. 4, pages 17-20.

Nichols, David B., 1993c, Open Celled Pavers: An Environmental Alternative to Traditional Paving Materials, pages 251-253 of *Proceedings of the 1993 Georgia Water Resources Conference*, Kathryn J. Hatcher, editor, Athens: University of Georgia Institute of Natural Resources.

Nightingale, Harry I., and William C. Bianchi, 1977, *Environmental Aspects of Water Spreading for Ground-Water Recharge*, Technical Bulletin 1568, Washington, D.C.: U.S. Department of Agriculture.

Oertel, Bob, 1993, Wildflowers for Stormwater Detention Basins, *Land and Water* January/February 1993, pages 26-27.

Paul Thiel Associates Limited, 1980, Subsurface Disposal of Storm Water, pages 175-193 of *Modern Sewer Design*, Washington: American Iron and Steel Institute.

Prill, Robert C., and David A. Aronson, 1978, *Ponding-Test Procedure for Assessing the Infiltration Capacity of Stormwater Basins, Nassau County, New York*, Water-Supply Paper 2049, Washington, D.C.: U.S. Geological Survey.

Rice, J.S., and B.W. Pinkerton, 1993, Reed Canarygrass Survival Under Cyclic Inundation, *Journal of Soil and Water Conservation* vol. 48, no. 2, pages 132-135.

Sorvig, Kim, 1993, Porous Paving, *Landscape Architecture* vol. 83, no. 2, pages 66-69.

Southerland, Robert J., 1984, Concrete Grid Pavers, *Landscape Architecture* vol. 74, no. 2, pages 97-99.

Strom, Steven, and Kurt Nathan, 1993, *Site Engineering for Landscape Architects*, second edition, New York: Van Nostrand Reinhold.

Thelen, Edmund, and L. Fielding Howe, 1978, *Porous Pavement*, Philadelphia: Franklin Institute Press.

van Hylckama, T.E.A., 1979, Changes in Vegetation Diversity Caused by Artificial Recharge, *Vegetatio* vol. 39, no. 1, pages 53-57.

Watshke, Thomas L., and Ralph O. Mumma, 1989, *The Effect of Nutrients and Pesticides Applied to Turf on the Quality of Runoff and Percolating Water*, ER 8904, University Park: Pennsylvania State University Environmental Resources Research Institute.

Weaver, Robert J., 1969, *Recharge Basins for Disposal of Highway Storm Drainage: Theory, Design Procedure, and Recommended Engineering Practices*, Research Report 69-2, Albany: New York State Department of Transportation, Engineering Research and Development Bureau.

Wells, Malcolm B., 1971, The Absolutely Constant Incontestably Stable Architectural Value Scale, *Progressive Architecture* vol. 52, no. 4, pages 92-97.

Wells, Malcolm B., 1974a, An Underground Office, *Progressive Architecture* vol. 55, no. 6, pages 112-113.

Wells, Malcolm B., 1974b, Environmental Impact, *Progressive Architecture* vol. 55, no. 6, pages 59-63.

Wells, Malcolm B., 1981, *Gentle Architecture*, New York: McGraw Hill.

Wingerter, Roger, and John E. Paine, 1989, *Field Performance Investigation, Portland Cement Pervious Pavement*, Orlando: Florida Concrete and Products Association.

3

Infiltration Hydrology

Infiltration is the movement of water from the surface into the ground, whether naturally occurring or artificially induced. Quantitative modeling of infiltration and its results is a basis for designing the hydrologic capacities of infiltration surfaces and basins on specific sites.

INFILTRATION PROCESS

Darcy's law

A basic principle governing flow in homogeneous porous media was formulated by Henri Darcy in 1856. Darcy's law describes a flow of constant rate in the form (Chow, Maidment and Mays, 1988, p. 39; Heath, 1983, p. 12)

$$Q = KA(\Delta h/L)$$

where

Q = flow, cm^3/h;

K = saturated hydraulic conductivity, a characteristic of a given porous medium when effectively saturated with water, cm/h;

A = cross-sectional area through the porous medium perpendicular to the flow, cm^2;

$\Delta h/L$ = hydraulic gradient, the difference in hydraulic head (Δh) per unit distance in the direction of flow (L), cm/cm.

Darcy's equation can be rewritten by substituting the equivalency $Q = qA$ to derive

$$q = K(\Delta h/L)$$

where

q = velocity of water through a unit cross section of the porous medium, called the darcian velocity, cm/h.

The velocity of fluid water through the pores of the medium is given by

$$V = q/\theta_s$$

where

V = fluid velocity, cm/h;

θ_s = water content of the medium, cm^3/cm^3, equal to the medium's porosity less the volume of trapped air in the medium's pores.

The hydraulic gradient $\Delta h/L$ is the driving force that makes water move. Where there is a difference in the total head between two points in a soil mass, water moves from the higher to the lower head, always moving in the direction of steepest gradient.

Hydraulic conductivity K is high in materials with large, continuous pores. In homogeneous granular soils, the highest conductivities are in soils composed of a single large grain size; any gradation in grain size would fill the pores with smaller particles. In addition soil is capable of developing structure—the aggregation of grains into larger particles or units. A soil with a high fraction of clay can be highly conductive if it has an aggregate structure, or it can be nearly impermeable if the particles are kept dispersed and structureless.

Infiltration

Infiltration into a homogeneous porous material follows Darcy's law. The application, when fully articulated, is complex because hydraulic gradient varies over time as ponding depth varies and as the matric suction is altered by added water. The use of computers makes complex applications feasible.

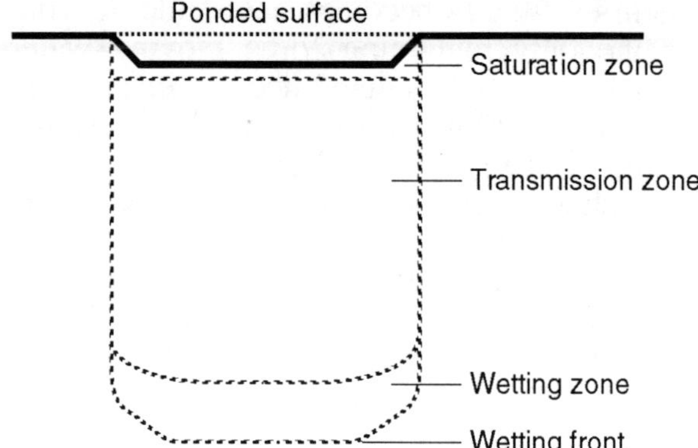

Figure 3.1 Moisture zones in the soil beneath a ponded surface.

The infiltration rate is the flux of water into the soil, in units of cm/h (Hillel, 1980, p. 5-49). When the rate of delivery of water to the surface is smaller than the soil's ability to take it in, water infiltrates as fast as it arrives. The infiltration rate rises when water is ponded over the surface, increasing the head. Horton (1940) hypothesized that the potential infiltration rate is controlled at the soil surface, but other workers have allowed for the possibility that hydraulic gradient, and thus infiltration rate, might be affected by conditions deep within the profile.

Wetting of a homogeneous soil under a ponded surface is shown in Figure 3.1 (Chow, Maidment and Mays, 1988, p. 108; Hillel, 1980). Normally only a zone near the surface is saturated, perhaps to a depth of several millimeters or centimeters. Beneath this zone is a lengthening transmission zone of unsaturated flow and fairly uniform moisture content. Beyond the transmission zone is a wetting zone, where moisture decreases with depth down to a wetting front, where the change of moisture content with depth is so great as to give the appearance of a sharp discontinuity between dry soil beneath and wet soil above.

Infiltration downward into an initially dry soil occurs under the combined influence of ponding head and suction gradient (Hillel, 1980). The suction gradient at first can be much greater than the gravity gradient resulting from ponding head. As the water penetrates deeper and as the transmission zone lengthens, the suction gradient decreases, because the difference in matric suction (between the saturated soil surface and the unwetted soil below the wetting front) divides itself along an increasing distance.

The suction gradient eventually becomes negligible, leaving gravity gradient as the only remaining force moving water downward. In vertical flow, each unit of decline in ponding depth (L) leads to an equal unit loss of gravity head (Δh), so the gravity gradient has the value of unity. Early in a ponding event the total hydraulic gradient is higher than unity, because the suction gradient and the gravity gradient are both significant. Over time the total hydraulic gradient declines, approaching a lower limiting value of 1.0, which is the lowest value that should be expected at any time in vertical infiltration (Miller and Richard, 1952). As long as the profile is homogeneous and structurally stable, vertical infiltration settles down to a steady, gravity-induced rate approaching the hydraulic conductivity as a lower limiting value. Thus 1.0 is the lowest value of total hydraulic gradient, and K is the lowest value of infiltration rate, that should be expected at any time in vertical infiltration from a ponded surface. In horizontal flow, the infiltration rate declines and approaches zero.

When surface ponding ceases, drainage through the vadose zone continues. In California (Bianchi and Muckel, 1970, p. 22-23), two experimental ponds were used to bring a 30 m depth of vadose zone up to a maximum moisture content by infiltrating 13,000 m^3 of water. After entry into the vadose zone, about half of the infiltrated water remained in the vadose zone after 100 days, and the total quantity of moisture took more than two years to drain entirely into the water table.

The Green-Ampt model

Chow, Maidment and Mays (1988, p. 109-117) and Hillel (1980, p. 12-13) reviewed mathematical models of infiltration, including those of Green and Ampt (1911), Holtan (1961), Horton (1940) and Philip (1957).

That of Green and Ampt (1911) is an example of methods having a useful combination of validity and simplicity (Hillel, 1980; Rubin, Glass and Hunt, 1976). It is based on a physical theory that is approximate, but its results are acceptably accurate (Chow, Maidment and Mays, 1988, p. 110). It illustrates the relationships that may constrain the capacities of infiltration surfaces and basins. It has been applied in the field to problems of infiltration into the vadose zone. It yields estimates of the infiltration rate and the cumulative infiltration depth at any given time during an infiltration event.

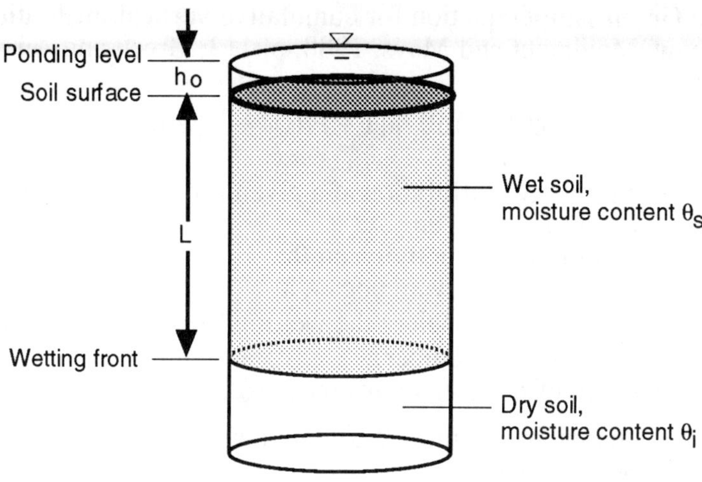

Figure 3.2 Infiltration into a column of soil of unit cross-sectional area as conceived in the Green-Ampt model (after Chow, Maidment and Mays, 1988, Figure 4.3.2).

The principal assumptions of the Green-Ampt approach are that the soil is homogeneous and stable, the supply of ponded water at the surface is not limiting, there exists a distinct and precisely definable wetting front, the matric suction just below the wetting front is uniform throughout the profile and constant in time during the infiltration event, and above the wetting front the transmission zone is uniformly wet (Hillel, 1980).

Consider a vertical column of homogeneous soil of unit horizontal cross-sectional area as shown in Figure 3.2 (Chow, Maidment and Mays, 1988, p. 111-113). If before the infiltration event the soil was of moisture content θ_i (cm/cm) throughout its depth, the moisture content increases from θ_i to the field saturated water content θ_s as the wetting front passes. The increase in the water stored in the wetted zone between the surface and depth L is $L(\theta_s - \theta_i) = L\Delta\theta$. This quantity is equal to F, the cumulative depth of water infiltrated into the soil.

The Green-Ampt equation for cumulative vertical infiltration is (Chow, Maidment and Mays, 1988, p. 113; Green and Ampt, 1911)

$$F = Kt + (\psi + h_o)\Delta\theta \ln[1 + F/((\psi + h_o)\Delta\theta)]$$

where

F	=	cumulative infiltration at a given time during the infiltration event, cm;
K	=	hydraulic conductivity, cm/h;
t	=	time since the event began, h;
ψ	=	matric suction below the wetting front, cm (entered as a positive number);
h_o	=	depth of ponding above the surface, cm;
$\Delta\theta$	=	change in soil moisture content as the wetting front passes, cm/cm.

Iterative calculation is necessary to resolve the circular reference to F on the left and right sides of the equation.

Given a solution for cumulative infiltration F at a given time, the corresponding infiltration rate f at that time can be found by (Chow, Maidment and Mays 1988, p. 113)

$$f = K[((\psi - h_o)\Delta\theta)/F + 1]$$

Application of the Green-Ampt model requires estimates of the hydraulic conductivity K, the suction head below the wetting front ψ, and the change in soil moisture content $\Delta\theta$. Rawls, Brakensiek and Saxton (1982) and Rawls, Brakensiek and Miller (1983) determined the values listed in Table 3.1 from laboratory analysis of approximately 5,000 soil horizons in 1,200 soil profiles across the United States. The table shows the average for each soil texture. The numbers in parentheses are one standard deviation around the average.

The change in soil moisture as the wetting front passes, $\Delta\theta$, is found by $\Delta\theta = (1 - s_e)\theta_e$ (Chow, Maidment and Mays, 1988, p. 114), where s_e is effective saturation below the wetting front (cm/cm) and θ_e is the soil's effective porosity. A typical value of s_e is 0.3. Effective porosity θ_e is the soil's porosity minus the moisture content that remains after thorough drainage; thus it is equal to the greatest content of water that can be added to a dry soil (Rawls et al., 1982). Rawls, Brakensiek and Miller (1983) gave the values of θ_e listed in Table 3.1 for individual textures.

Table 3.1 Parameters in the Green-Ampt model (effective porosity and wetting front soil suction head from Rawls, Brakensiek and Miller, 1983; saturated hydraulic conductivity from Rawls, Brakensiek and Saxton, 1982).

Soil texture	Effective porosity θ_e (cm^3/cm^3)	Suction head ψ below wetting front (cm)	Saturated hydraulic conductivity K (cm/h)
Sand	0.417 (0.354-0.480)	4.95 (0.97-25.36)	21.00
Loamy sand	0.401 (0.329-0.473)	6.13 (1.35-27.94)	6.11
Sandy loam	0.412 (0.283-0.541)	11.01 (2.67-45.47)	2.59
Loam	0.434 (0.334-0.534)	8.89 (1.33-59.38)	1.32
Silt loam	0.486 (0.394-0.578)	16.68 (2.92-95.39)	0.68
Sandy clay loam	0.330 (0.235-0.425)	21.85 (4.42-108.0)	0.43
Clay loam	0.309 (0.279-0.501)	20.88 (4.79-91.10)	0.23
Silty clay loam	0.432 (0.347-0.517)	27.30 (5.67-131.50)	0.15
Sandy clay	0.321 (0.207-0.435)	23.90 (4.08-140.2)	0.12
Silty clay	0.423 (0.334-0.512)	29.22 (6.13-139.4)	0.09
Clay	0.385 (0.269-0.501)	31.63 (6.39-156.5)	0.06

The limitations of using data based on texture alone must be understood. The values experienced in the field can deviate from the averages listed in Table 3.1, as shown by the magnitudes of the standard deviations. In addition differences among soils due to structure and crusting can be greater than those due to texture. Standardized data such as those in Table 3.1 are necessary in design, but they must be supplemented following construction with site-specific monitoring and soil maintenance in order to assure that a basin functions properly.

Bouwer (1966) found that infiltration rate into a given dry soil is limited by air trapped in the soil voids, reducing the available porosity into which water can infiltrate. The rate of infiltration was about half the rate that would be predicted by the use of the saturated hydraulic conductivity. Thus for infiltration it is appropriate to introduce a safety factor Sf to the saturated hydraulic conductivity equal to about 0.5 and to substitute the product KSf for K in the Green-Ampt model. This was reflected by Rawls et al.'s (1983) publication of soil conductivity during sorption equal to half the saturated hydraulic conductivity for the same soil texture.

When water infiltrates into a layer of high conductivity overlying a less conductive layer, the infiltration rate is at first controlled by the upper layer. When the wetting front reaches the deeper, less conductive layer, the infiltration rate may drop, approaching that of the deeper soil alone. Chow, Maidment and Mays (1988, p. 116-117) developed upon the Green-Ampt model for infiltration into this type of layered soil. The original (single-layer) Green-Ampt equations are used while the wetting front is in the upper layer; the two-layer equations are used when the wetting front has penetrated through the upper layer and into the lower layer an additional depth L_2. The upper layer has thickness H_1 and Green-Ampt parameters K_1, ψ_1 and $\Delta\theta_1$, and the lower layer has K_2, ψ_2 and $\Delta\theta_2$. The cumulative infiltration is given by $F = H_1\Delta\theta_1 + L_2\Delta\theta_2$. The rate is given by $f = [(K_1K_2)/(H_1K_2 + L_2K_1)](\psi_2 + H_1 + L_2)$. If infiltration continues for a long time then a positive pressure head (a perched water table) can develop in the coarse soil.

In the opposite case of infiltration into a fine-textured layer, such as a surface crust, overlying a more conductive one, the initial infiltration rate is again determined by the upper layer. As water reaches the interface with the more conductive layer, the infiltration rate may decrease. This is because the suction in the fine-textured layer may be too great to permit entry into the relatively large pores of the coarse layer until the total head at the interface builds up sufficiently for water to penetrate into the coarse material. When it does, the lower layer is not saturated by infiltrating water, because the rate of flow through the less permeable layer is restricted (except when the externally applied ponding head is large). The percolating water may concentrate at points along the interface to break through in finger-like protrusions into the coarse-textured layer. Flow through the more conductive layer thereafter occurs through "chimneys" or "pipes." The overall flow rate through the system is governed by the number of fingers per unit area of interface.

Morel-Seytoux (1985) expanded on the Green-Ampt approach to describe redistribution of the moisture profile after ponding ceases, evolution of a ground water mound, and recharge from a basin in hydraulic (saturated) connection with the water table.

Variations in earth materials

The moisture and infiltration characteristics of earth materials vary greatly. Vertically, soils tend to evolve horizons or layers such that almost any soil characteristic can vary widely within a few centimeters. Horizontally, hillslopes, vegetation, depositional processes and land use history can impose variations within small distances.

The electrochemical forces at the surfaces of clay particles can bind particles together into larger aggregates, building macropores and increasing conductivity. On the other hand the clay particles in some soils are physicochemically dispersive, making the soil susceptible to disaggregation and formation of an unstructured, dense layer of low conductivity for a given texture.

Certain chemicals, most notably sodium, aggravate the dispersion of soil colloids (Jackura, 1980, p. 73). Sodium concentration in soil is high in some arid areas and where roads are salted to melt ice during the winter. Gypsum counteracts the dispersive effects of other chemicals, increasing aggregation, porosity and conductivity (Patterson, 1976).

Organic matter can assist clay particles in building soil structure by gluing particles together. In a drained, aerobic soil any addition of organic material or any maintenance practice that preserves leaves and other naturally produced organic matter improves structure and porosity. There must be a continuing supply of fresh organic matter to maintain infiltration rates as organic material decomposes (McCalla, 1945).

The development of a macropore system represents a balance between constructive and destructive processes (Beven and Germann, 1982). Any change in the soil-plant-animal community and its external conditions, such as weather patterns or surface ponding, affects that balance. In vegetated soil, macroporosity is continuously regenerated by plants and animals. The most effective macropore systems occur in undisturbed, usually forested, soils. The lowest macroporosity occurs where soil disturbance and compaction are most intense. In recently excavated or disturbed soils, time is required for macropore establishment. Under favorable conditions macropore systems may be produced within one or two years. The effective lifetime of macropores increases with stability of the soil structure. In mature soils, some macropores may be hundreds of years old.

Macropores occur mostly in the upper horizons, where they can occupy 35 percent or more of the volume of a maturely forested soil. Many macropores are readily visible and may be continuous for several meters vertically and horizontally (Beven and Germann, 1982). Macropores formed by plant roots are frequently filled with loosely packed organic matter from the decaying roots. Large hollows formed by decaying tree stumps branch into networks of linear pores; grass roots form more equally sized pores. Soil fauna such as earthworms, insects, ants, moles and gophers form macropores ranging in diameter from less than 1 mm to over 50 mm. Other macropores are formed by soil shrinkage during desiccation of clays, chemical weathering of bedrock, freeze-thaw cycles, and erosion by subsurface flows through noncohesive materials.

Flow through a soil layer containing macropores cannot be accurately described by approaches based on Darcy's law, because Darcy's assumption of a homogeneous soil-pore matrix does not apply (Beven and Germann, 1982). The soil mass is permeated by channels which function as major hydraulic pathways, even though they may contribute only a small amount to total porosity. Macropores can conduct water through unsaturated soil ahead of the wetting front in the matrix, bypassing most of the soil volume. In soils of low to moderate matrix conductivity, even a small amount of macroporosity can increase hydraulic conductivity by more than 10 times; clay soils can acquire conductivities usually associated with gravels.

INFILTRATION RESULTS

After water from a rainfall or snowmelt enters the soil, the potential quantity of direct surface runoff is reduced and the water undergoes instead the longer, more continuous processes of evapotranspiration, ground water recharge, and base flow. One of the purposes of stormwater infiltration is to influence the distribution of water among these processes so that a site's hydrology can support environmental systems and resources from season to season and year to year.

Quantitative modeling of a site's water balance is a basis for evaluating the effects of alternative site development and stormwater design solutions. It can also be used to estimate the runoff inflows to an infiltration basin for periods longer than an individual storm. A computation of the water balance produces a continuous record of soil moisture, evapotranspiration and runoff, based on input of a meteorological record and certain characteristics of soil, vegetation and land use.

The long-term water balance

The long-term water balance refers specifically to the average levels and seasonal fluctuations of storages and flows in an average year. It has long been a prominent concept in geography, where it is used as a summary index of the moisture and energy endowments of regional environments.

An early approach to water balance computation was that of Thornthwaite (Thornthwaite and Mather, 1955), who developed a "bookkeeping" procedure for partitioning monthly precipitation into evapotranspiration and runoff, and for carrying soil moisture and other varying stores of water from month to month. The concept is shown in Figure 3.3, where soil moisture and detention are storages, and the arrows represent flows. Some precipitation is deflected into direct runoff (storm flows) depending on soil cover and rainfall intensity (in cold regions, some precipitation is also deflected into snowpack storage). The remaining water enters the soil root zone. From the root zone some of it is lost to evapotranspiration, the amount depending on the balance between climatic reference (potential) evapotranspiration and matric suction. Any further remaining water, referred to by Thornthwaite as monthly surplus, percolates below the root zone, into the ground water, from which it drains slowly as base flow. The total runoff is the sum of direct runoff and base flow.

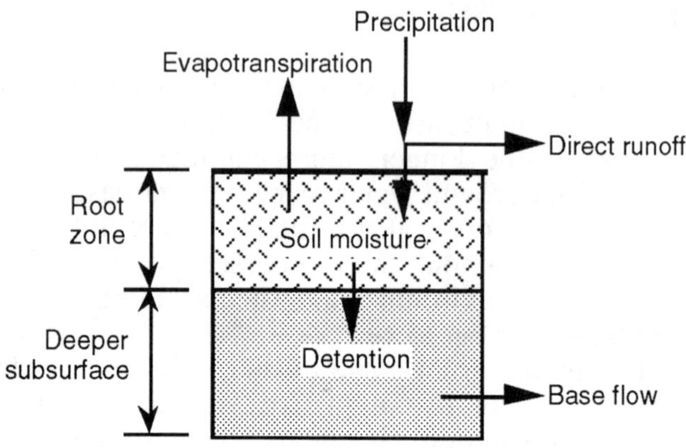

Figure 3.3 Storages and flows in a soil profile as conceived in the Thornthwaite water balance (Thornthwaite and Mather, 1955).

The use of average monthly data to determine the water balance requires only moderate climatic data input and computing capabilities. Perhaps, in the future, evolutions in hardware and accessibility of shorter-term data will make the use of average monthly data obsolete. In the meantime, Thornthwaite's concept of the water balance is useful for understanding the types of effects that land use and stormwater management choices can have on site environments. His concept also provides a basis for understanding more sophisticated models when they become more widely used.

The data in Table 3.2 were generated this way for an urban site in the Chicago area. Some important results are graphed in Figure 3.4. It can be seen that at this location precipitation occurs in all months of the year, reaching a maximum in the summer. Reference (potential) evapotranspiration varies more widely, from zero in the winter to higher than precipitation in the summer, creating a climatic moisture surplus in the winter and a deficit in the summer. Soil moisture, ground water and base flow are all high in the winter and early spring; this is a time of relative abundance for vegetation and aquatic life. Direct runoff is briefly high in the early spring, even higher than monthly precipitation, while snow is melting and soil moisture is high. Direct runoff stays relatively high during the rainy summer due to the soil's partly impervious surface and high urban curve number. During this time flash floods and channel

Table 3.2 Water balance for an urban site in the Chicago area. All values are in mm except number of precipitation days and SCS storage factor S. Mean meteorologic data for 1941-1970 from the National Oceanic and Atmospheric Administration. Et_o was estimated by the Thornthwaite method (Thornthwaite and Mather, 1957). Soil AWHC was assumed to be 112 mm based on weighted average of pervious and impervious areas. CN II was assumed to be 70, from which it was calculated that CN I = 50.6 and CN III = 85.3.

	Jan	Feb	Mar	Apr	May	Jun	Jul	Aug	Sep	Oct	Nov	Dec
Meteorologic data												
Precipitation P	47	40	69	95	87	100	104	80	76	67	56	54
Precipitation days D	11	10	13	13	12	11	9	8	9	8	10	11
Et_o	0	0	7	42	82	131	147	132	91	50	12	0
Snowpack												
Cumulative snowpack	10	31	41	30	0	0	0	0	0	0	0	0
Precipitation to snowpack	20	10	0	0	0	0	0	0	0	0	0	10
Snowpack melt	0	0	11	30	0	0	0	0	0	0	0	0
Net ΔSnowpack	20	10	-11	-30	0	0	0	0	0	0	0	10
Net precipitation	27	30	80	125	87	100	104	80	76	67	56	43
Direct runoff												
SCS storage S	2.0	0.9	2.0	0.9	5.9	5.9	5.9	5.9	5.9	5.9	5.9	0.9
Direct runoff DRo	4	11	58	102	55	53	51	31	25	21	10	7
Soil moisture & Et												
Net inflow to soil	22	19	15	-19	-50	-83	-94	-84	-39	-4	34	36
APWL	0	0	0	-19	-69	-153	-247	-331	-370	-374	0	0
Soil moisture SM	96	112	112	94	60	29	12	6	4	4	38	74
ΔSM	22	16	0	-18	-34	-32	-16	-6	-2	0	34	36
Evapotranspiration Et	0	0	7	41	66	79	69	55	53	46	12	0
Base flow												
Moisture surplus	0	3	15	0	0	0	0	0	0	0	0	0
Available for BRo	0	3	16	8	4	2	1	1	0	0	0	0
Base flow BRo	0	2	8	4	2	1	1	0	0	0	0	0
Detention	0	2	8	4	2	1	1	0	0	0	0	0

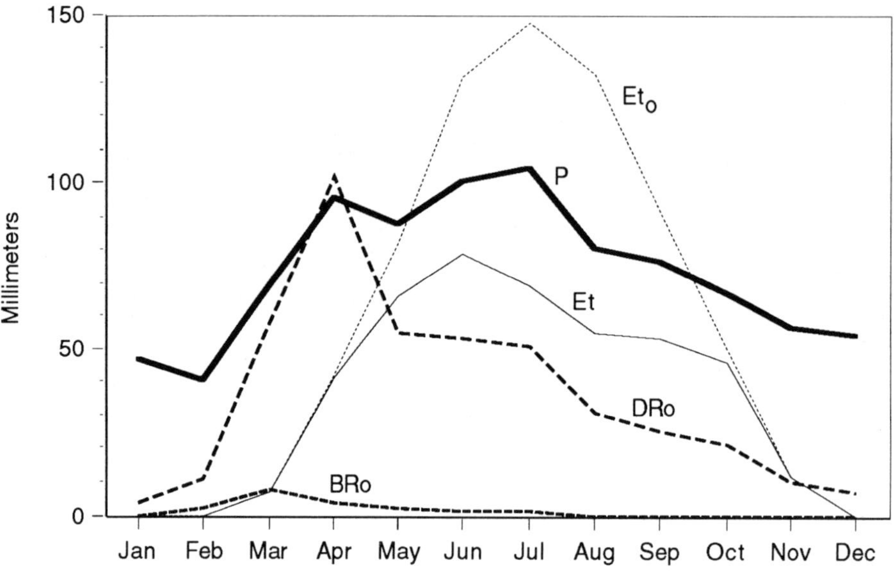

Figure 3.4 Long-term water balance for an urban site in the Chicago area (data from Table 3.2). P is precipitation; Et_o is reference evapotranspiration; Et is actual evapotranspiration; DRo is direct runoff; BRo is base flow.

erosion are possible. Base flow is low the rest of the year due to the loss of water as direct runoff from impervious surfaces. Actual evapotranspiration—that able to occur within the limitation of available soil moisture—rises in the summer, drawing down the previously stored soil moisture. High evapotranspiration indicates relatively high productivity in both crops and natural ecosystems. Nevertheless, the remaining deficit between Et_o and Et indicates potential plant stress. If more water were made available in the root zone, rather than being shunted off as direct runoff, plant stress would not be as great. In the late summer and fall, all runoff is low because the root zone is dry and ground water was never significantly replenished; this is a time of stress for stream flow, aquatic life, recreation, and human water supplies. Moisture surplus is restored in December, when evapotranspirative potential declines and the winter's snowpack begins to accumulate.

Calculations for Thornthwaite's method were reviewed in Thornthwaite and Mather's (1957) original manual and by Dunne and Leopold (1978, p. 238-248). The calculations are easily set up in spreadsheet form, with the columns and rows arranged like those in Table 3.2. The calculations for an average year are iterative, cycling from the end of December back into January and among circular references within each month.

The calculations are reviewed below in some detail because they are seldom covered in the hydrologic textbooks addressed to designers. The following review reflects adaptations by Ferguson (1990) and Ferguson, Ellington and Gonnsen (1991) specifically for simulating the effects of urbanization and alternative methods of stormwater control. All flows are in units of mm/month. While studying the following calculations, the reader can refer to Table 3.2 as an example of the completed results.

Meteorologic data

Basic regional monthly meteorologic data necessary for water balance computations are readily available. Data for all stations in a state are published in the periodical *Climatological Data* (U.S. National Oceanic and Atmospheric Administration, monthly and annual). For specific stations data are published in *Local Climatological Data* (U.S. National Oceanic and Atmospheric Administration, monthly and annual). These publications are available from the National Oceanic and Atmospheric Administration's National Climatic Data Center (Federal Building, Asheville, North Carolina 28801). The Center can be contacted to determine availability of detailed reports or printouts for specialized purposes. The Center's basic data are reprinted commercially by the Gale Research Company (multiple dates) in *The Weather Almanac* and *Weather of U.S. Cities*. The data are also available in digital form from commercial suppliers.

Evapotranspiration is not monitored as commonly as other meteorologic factors, nor are the data published as widely. Reference evapotranspiration can be estimated from other meteorologic data by a variety of physically based methods with widely varying complexity and accuracy (Dunne and Leopold, 1978; Ferguson, 1990; Jensen, 1990). One is the Thornthwaite method (Thornthwaite and Mather, 1957), which is based on simple inputs of temperature and latitude. A much more rigorous method, much more demanding of data, is that of Penman (1948).

Alternatively, reference evapotranspiration can be estimated from measurements in Class A evaporation pans, which measure evaporation more or less directly but under peculiar standard conditions. The measured data can be multiplied by an appropriate coefficient to convert to reference evapotranspiration from the land or water surface (Dunne and Leopold, 1978). The most common value of the coefficient is 0.7. For very small ponds the coefficient may be higher, sometimes over 0.9. In some regions the coefficient varies widely over the year. The U.S. Weather Bureau's (1964) *Climatography of the United States No. 86* listed mean monthly pan evaporation for monitored stations through the early 1960s. Issues of this publication for individual states are available from the National Climatic Data Center. The data for some some stations are updated in *Climatological Data* (National Oceanic and Atmospheric Administration, monthly and annual). For an individual station where pan evaporation is measured, copies of the original forms used to report the data to NOAA, *Form WSE-22*, are available from the Center. Woolhiser and Wallace (1984) produced maps of statistical variables sufficient to predict monthly pan evaporation in the United States east of the Rocky Mountains.

In certain locales measured or estimated evapotranspiration data are available from local agencies. In California, the California Department of Water Resources makes pan data from 80 stations available through the California Irrigation Management System. Water authorities in the Phoenix-Tucson area apply the Penman estimate to data from about 15 monitoring stations. Similar evapotranspiration data are available from water authorities in the Denver area.

Snowpack

In regions where snowpack stores a significant proportion of precipitation from month to month, accumulated snow is taken into account as a kind of reservoir. Inflow to the snowpack is freezing precipitation; outflow is melting snow. In the following equations P is precipitation in mm, T is monthly temperature in °C, and E is a factor arbitrarily assumed equal to 0.1.

perature in °C, and E is a factor arbitrarily assumed equal to 0.1.

P added to snow = EP(0 - T), if $T < 0$;

P added to snow = 0, if $T \geq 0$;

Snowmelt = ET(cumulative snowpack depth), if $T \geq 0$;

Snowmelt = 0, if $T < 0$.

The change in snowpack during a month is inflow minus outflow. The cumulative snowpack depth is equal to the previous month's snowpack plus the change in snowpack.

Where snowpack routing is taken into account in a water balance, the net available precipitation after routing replaces precipitation P in all the calculations that follow. It is given by

Net available P = P - P added to snow + snowmelt.

Direct runoff

Direct runoff estimation was considered by Thornthwaite and Mather (1957) to be only a refinement of the water balance, but its inclusion is mandatory for evaluation of urban development and design of stormwater infiltration. Direct runoff is the only runoff inflow to an infiltration basin in a small urban drainage area, and its capture and transformation is one of the purposes of infiltration design.

One way to estimate monthly direct runoff is to assume that it is a fixed ratio to precipitation. This is valid where most of the drainage area is impervious, so that the high runoff/precipitation ratio does not significantly vary with soil moisture in the small pervious area. The value of the ratio could be chosen based on the runoff coefficient in the rational formula, or the ratio predicted by the SCS method (U.S. Soil Conservation Service, 1986) during a small (two year or less) storm.

Another way to estimate direct runoff is to vary the runoff-to-precipitation relationship with seasonally varying soil moisture. This can be done with an adaptation of the SCS method.

The SCS method is based on curve numbers relating runoff to rainfall for specific soil-land use complexes. The curve numbers presented in manuals such as *Technical Release 55* (U.S. Soil Conservation Service, 1986) apply to average soil moisture conditions (CN II). The curve number of a given soil-land use complex varies from time to time from dry conditions (CN I) to wet conditions (CN III). Curve numbers not listed in standard manuals can be computed by (Chow, Maidment and Mays, 1988, p. 149):

$$\text{CN I} = 4.2\ \text{CN II} / (10 - 0.058\ \text{CN II})$$

$$\text{CN III} = 23\ \text{CN II} / (10 + 0.13\ \text{CN II})$$

The SCS curve number is related to soil moisture by the storage factor S, assumed proportional to the unfilled portion of the soil's available water holding capacity AWHC. A soil's minimum storage, S_{min}, occurs in wet conditions; it is 1000/CN III - 10. The maximum storage, S_{max}, occurs in dry conditions; it is 1000/CN I - 10. Where SM is soil moisture in mm and AWHC is available water-holding capacity of the soil root zone, the value of S in a given month can be found by

$$S = S_{min} + (S_{max} - S_{min})(AWHC - SM)/AWHC.$$

Table 3.3 lists available water holding capacities of soils on the basis of texture and rooting depth. According to the table, the available water holding capacity of a silt loam is 0.197 mm/mm of depth, or 19.7 percent by volume. The depth of rooting under the prevailing vegetation determines the depth of the rooting zone and thus the total depth of available water holding capacity. For a silt loam with moderately deeply rooted vegetation, the AWHC is (0.197)(1.00 m) = 197 mm.

Given the soil moisture level in a given month, S can be computed from the above formula, CN can be computed from S, and direct runoff DRo can be computed from the curve number and the monthly precipitation.

However, the SCS method was designed to be used with daily precipitation, not monthly. This can be taken into account by assuming that a month's total precipitation is equally distributed among the monthly number of days with precipitation. The original SCS equation for direct runoff DRo is given by (U.S. Soil Conservation Service, 1986):

$$DRo = ((P - 0.2S)^2)/(P + 0.8S), \text{ if } P > 0.2S;$$
$$DRo = 0, \text{ if } P \leq 0.2S.$$

Applying the SCS equation to precipitation days only, then multiplying times the monthly number of precipitation days D gives:

$$DRo = D((P/D - 0.2S)^2)/(P/D + 0.8S), \text{ if } P/D > 0.2S;$$
$$DRo = 0, \text{ if } P/D \leq 0.2S.$$

Soil moisture and evapotranspiration

Net inflow of moisture to the soil is that remaining after DRo and reference (potential) evapotranspiration Et_o:

$$\text{Net inflow to soil} = P - DRo - Et_o.$$

Net inflow may be either positive or negative; its sign defines meteorologically wet (+) and dry (-) months. The severity of the consecutive dry season demanding evapotranspiration of soil moisture is indicated by the accumulated potential water loss APWL. It is found as the cumulation of negative values of net inflow (to an arbitrary limit such as -9,999 necessary for computation of desert water balances, which would otherwise iterate forever toward infinitely high APWL).

If net inflow < 0:

APWL = previous APWL + current month's net inflow.

If net inflow > 0, and next net inflow < 0, and SM < AWHC:

APWL = AWHC(ln(SM/AWHC)).

If neither of the foregoing conditions are met:

APWL = 0.

Soil moisture SM is lost to evapotranspiration in accord with a balance between the meteorologic demand (the accumulated potential water loss) and the matric and gravity forces resisting withdrawal. In wet months, evapotranspiration is unrestrained by matric forces resulting from low soil moisture. In dry months, soil moisture responds to APWL according to the soil's AWHC and current moisture content. When soil moisture is high, it can be withdrawn freely; when low, it clings to the soil under matric suction.

If net inflow ≥ 0:

SM = previous SM + net inflow, to the limit of the AWHC.

If net inflow < 0:

SM = AWHC(exp(APWL/AWHC)).

The change in soil moisture ΔSM is the difference between the current month's SM and the previous SM.

Evapotranspiration Et can be figured from the month's reference (potential) evapotranspiration Et_o and the given change in soil moisture:

$Et = P - DRo - \Delta SM$, if $\Delta SM < 0$;

$Et = Et_o$, if $\Delta SM \geq 0$.

Base flow

Moisture surplus is the amount of moisture exceeding all current monthly demands and the soil's available water holding capacity. The surplus drains below the root zone, out of the reach of evapotranspiration, and is assumed equal to ground water recharge.

Surplus = net inflow - ΔSM,
if net inflow > 0 and ΔSM > 0;

Surplus = net inflow, if net inflow > 0 and ΔSM ≤ 0;

Surplus = 0, if net inflow ≤ 0.

Table 3.3 Available water holding capacity (AWHC) for various root zone textures and depths. (Data for impervious cover assumed. For vegetated soils, rooting depths from Thornthwaite and Mather, 1955, with depths at some textures interpolated. AWHC/depth derived from data on water retained at 1/3 bar and 15 bars tension in Rawls, Brakensiek and Saxton, 1982. AWHC of root zone derived by multiplication.)

Vegetation or cover	Soil texture	Rooting depth (m)	AWHC/depth (mm/mm)	AWHC of root zone (mm)
Impervious	Not applicable	0.001	1.000	1
Shallow rooted (spinach, peas, beans, beets, carrots)	Sand	0.50	0.058	29
	Loamy sand	0.50	0.070	35
	Sandy loam	0.50	0.112	56
	Loam	0.56	0.153	86
	Silt loam	0.62	0.197	122
	Sandy clay loam	0.51	0.107	55
	Clay loam	0.40	0.121	48
	Silty clay loam	0.36	0.158	57
	Sandy clay	0.33	0.100	33
	Silty clay	0.29	0.137	40
	Clay	0.25	0.124	31
Moderately deep rooted (corn, cereals, cotton, tobacco)	Sand	0.75	0.058	44
	Loamy sand	0.88	0.070	62
	Sandy loam	1.00	0.112	112
	Loam	1.00	0.153	153
	Silt loam	1.00	0.197	197
	Sandy clay loam	0.90	0.107	96
	Clay loam	0.80	0.121	97
	Silty clay loam	0.73	0.158	115
	Sandy clay	0.65	0.100	65
	Silty clay	0.58	0.137	79
	Clay	0.50	0.124	62
Deeply rooted (alfalfa, shrubs, pasture grass)	Sand	1.00	0.058	58
	Loamy sand	1.00	0.070	70
	Sandy loam	1.00	0.112	112
	Loam	1.13	0.153	173
	Silt loam	1.25	0.197	246
	Sandy clay loam	1.13	0.107	121
	Clay loam	1.00	0.121	121
	Silty clay loam	0.92	0.158	145
	Sandy clay	0.84	0.100	84
	Silty clay	0.75	0.137	103
	Clay	0.67	0.124	83

Table 3.3 (continued).

Vegetation or cover	Soil texture	Rooting depth (m)	AWHC/depth (mm/mm)	AWHC of root zone (mm)
Orchards	Sand	1.50	0.058	87
	Loamy sand	1.59	0.070	111
	Sandy loam	1.67	0.112	187
	Loam	1.59	0.153	243
	Silt loam	1.50	0.197	296
	Sandy clay loam	1.25	0.107	134
	Clay loam	1.00	0.121	121
	Silty clay loam	0.92	0.158	145
	Sandy clay	0.84	0.100	84
	Silty clay	0.75	0.137	103
	Clay	0.67	0.124	83
Mature forest	Sand	2.50	0.058	145
	Loamy sand	2.25	0.070	158
	Sandy loam	2.00	0.112	224
	Loam	2.00	0.153	306
	Silt loam	2.00	0.197	394
	Sandy clay loam	1.80	0.107	193
	Clay loam	1.60	0.121	194
	Silty clay loam	1.49	0.158	235
	Sandy clay	1.39	0.100	139
	Silty clay	1.28	0.137	175
	Clay	1.17	0.124	145

Base flow is the discharge of stored moisture surplus from the ground water. Thornthwaite and Mather (1957) assumed that half of the ground water discharged as BRo each month, leaving the other half in storage to carry into the following month (the storage is "detention" in Thornthwaite and Mather's original phrase):

Available for BRo = surplus + previous month's detention
BRo = 0.5(available for BRo)
Detention = 0.5(available for BRo)

INFILTRATION BASIN THROUGHFLOWS

Water that enters an infiltration basin is partitioned over time into infiltration, evaporation, overflow and change in basin storage. Thus an infiltration basin has its own water balance. Quantitative modeling of a basin's throughflows and storage is a basis for infiltration basin design, whether during individual storm events or over the seasons in the long-term water balance. It is similar in principle to any other reservoir routing.

The inflow to a basin is direct runoff from the drainage area and, if the basin is open at the surface, direct precipitation. Direct runoff (m^3) entering a basin during a storm event can be estimated using any one of a variety of documented rainfall-runoff models. They all take into account the size and condition of the drainage area and characteristics of rainfall. Some of them were discussed by Ferguson and Debo (1990), Chow, Maidment and Mays (1988) and the U.S. Soil Conservation Service (1986). Published sources of storm precipitation data to use as input to the models include the U.S. Weather Bureau's *Technical Paper 40* (Hershfield, 1961), the U.S. Soil Conservation Service's (1986) *Technical Release 55*, and local drainage manuals.

Ponding time

Ponding time is a simple summary way to characterize a basin's behavior during a storm event. If a given basin is assumed to be filled to capacity instantaneously, with no further inflows during the event, the given depth of ponded water will take a certain amount of time to infiltrate. The volume infiltrated during this time is depth of water (in a subsurface basin, total depth times void space) times basin floor area.

One approach to estimation of ponding time is to assume that the hydraulic gradient is constant at 1.0 and thus that the infiltration rate is a constant equal to the soil's hydraulic conductivity K. Thus, with all infiltration through a level floor, ponding time = depth/K. With an appropriate safety factor Sf applied to the conductivity, ponding time = depth/(KSf). In Figure 3.5, one of the curves shows cumulative infiltration through the floor of a 100-m^2 basin at a constant K of 0.68 cm/h. With an initial ponding depth of 75 cm, total ponding time by this method is 110 h.

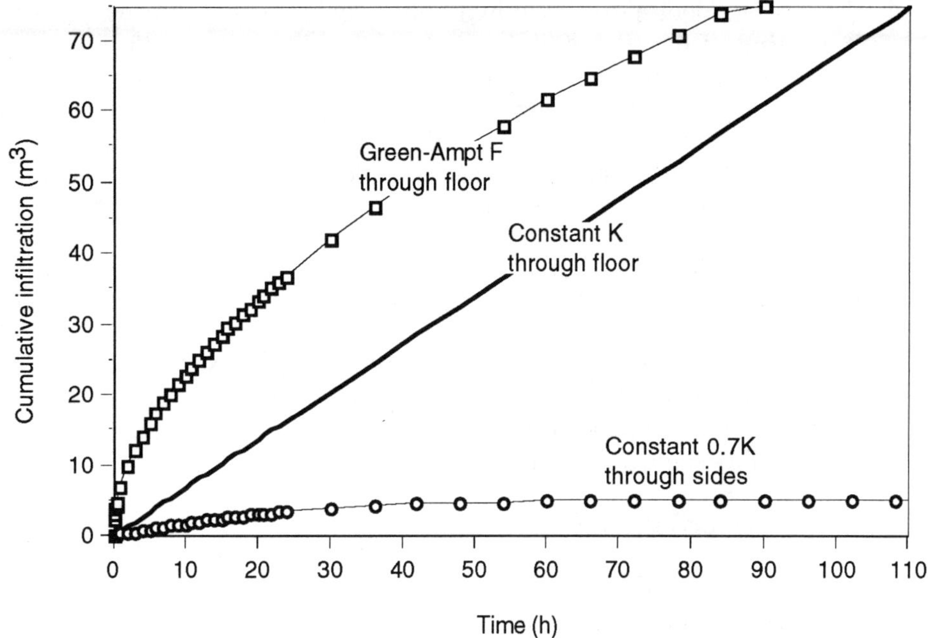

Figure 3.5 Cumulative infiltration from a hypothetical basin predicted by the constant-rate and Green-Ampt methods. The basin is initially ponded 75 cm deep; thus the initial ponded volume is 75 m^3. K is 0.68 cm/h.

An alternative approach is to apply the Green-Ampt model. The depth of ponding at the beginning of the event is the initial h_o. The Green-Ampt equation can be applied repetitively to track the gradual decline of the ponded water surface toward the basin floor. The depth of ponding in each time increment is equal to the initial depth h_o less the cumulative infiltration. The entire ponded volume has infiltrated when cumulative infiltration F equals the initial ponded depth.

One of the curves in Figure 3.5 shows ponding behavior estimated with a spreadsheet application of Green and Ampt's equation for F. Data entered for this example were $\psi = 16.68$ cm, $\theta_e = 0.486$, and $s_e = 0.3$. The initially steep slope of the F curve shows that the Green-Ampt infiltration is rapid at the beginning of the event, when total hydraulic gradient is high. The slope declines as total hydraulic gradient is reduced by the wetting of the underlying soil. Total ponding time is 85 h. By this method, cumulative infiltration is greater at any time than with the constant-rate approach, and total ponding time is shorter due to the relatively rapid infiltration early in the event.

Infiltration through a basin's sides can be taken into account by assuming that the direction of flow into the soil is 45° from the vertical, in which case a constant gravity gradient would be 0.7. The wetted area of side surface through which the water infiltrates can be approximated at any time as the stage times the basin's perimeter length. The third curve in Figure 3.5 shows an application of this approach to the same basin, given a perimeter length of 50 m. Some infiltration occurs early in the event when most of the side area is wet, but quickly levels off as stage and wetted side area decline.

Overflow rate

During a storm an infiltration basin captures a given volume of runoff, beginning with the first runoff during the storm event. Runoff in excess of the basin's capacity overflows the basin and continues as surface drainage downstream. When overflow occurs, its rate is equal to the rate of inflow, unless surface detention provisions have been made.

Figure 3.6 shows the rate of flow Q arriving at a basin from a drainage area as a function of the cumulative flow volume during the event, according a standard unit hydrograph shape presented by the Soil Conservation Service (1972, Table 16.1). With Q_{vol} being the total flow volume during the event, Q is low early in the event, peaks at $0.375Q_{vol}$, and then declines. If a basin captures a volume equal to $0.375Q_{vol}$ or less, and all remaining runoff flows over the basin as fast as it arrives, then the peak rate of overflow equals the peak rate of inflow Q_p. If a basin captures more than $0.375Q_{vol}$, Q_p is absorbed by the basin, and the peak rate of overflow is a point on the declining limb of the graph. For instance, if the basin captures a volume of $0.7Q_{vol}$, the peak rate of overflow is $0.68Q_p$. In this case, the overflow's time to peak will be longer than that of the inflow.

Other specific hydrograph shapes can be estimated for given site designs. Nevertheless, it is always true that, if an infiltration basin captures a sufficient volume of runoff during a given storm, then it reduces the peak flow rate at the point of discharge.

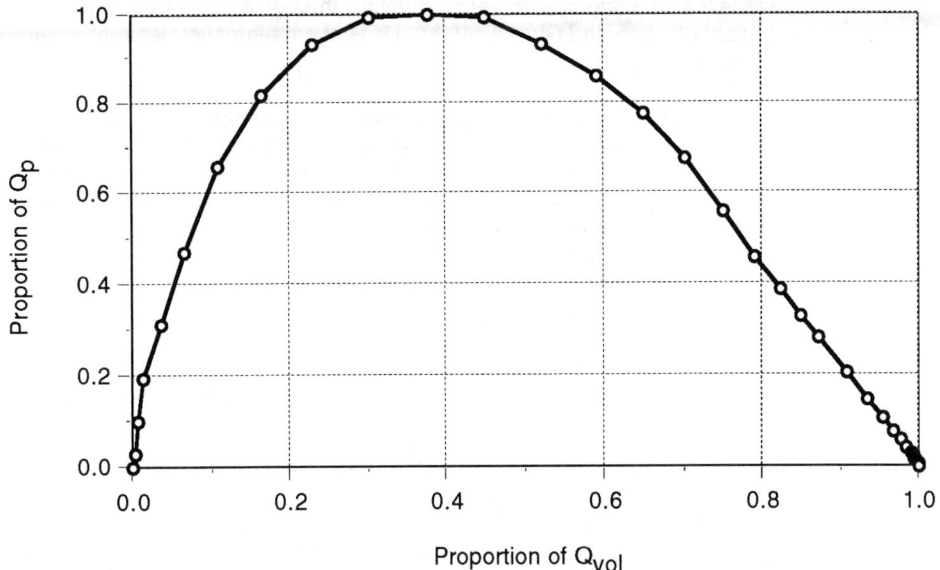

Figure 3.6 Rate of runoff during a storm as a function of cumulative runoff volume (data from U.S. Soil Conservation Service, 1972, Table 16.1). Q_{vol} is total runoff volume during the event (m^3); Q_p is peak rate of runoff (m^3/s).

Ponding with varying inflow

When a basin's varying inflows over time are taken into account, the change in amount of water in storage during any time period is a balance between inflow and outflow.

The inflow hydrograph is specified by entering the inflow volume in each time increment. If storage is in units of depth of water in the basin, then inflow volume is converted to depth by dividing by the basin's floor area.

The basin's outflow in each time increment is the sum of infiltration plus surface overflow (if stage exceeds the depth of the basin) plus evapotranspiration (if the basin is exposed at the surface).

Storage depth at any time is the difference between cumulative inflow and cumulative outflow. For a subsurface basin, depth of water can be converted to total basin depth (stage) by dividing by the basin's void space. Depth of stored water can be converted to storage volume by multiplying by the basin's floor area.

Table 3.4 Routing of a 25-year storm flow through a basin near Philadelphia (Urban and Gburek, 1980). Units of cm represent depth of water in the basin.

Time (h)	Precipitation intensity (cm/h)	Cumulative precipitation (cm)	Cumulative infiltration (cm)	Storage (cm)
0.5	9.91	4.95	0.16	4.83
1.0	6.35	6.35	0.33	6.10
2.0	3.86	7.75	0.66	7.11
3.0	2.87	8.64	0.99	7.62
6.0	1.75	10.41	1.98	8.38
12.0	1.07	12.70	3.96	8.89
24.0	0.58	14.22	7.92	6.10

A simple illustration is a routing of a 25-year storm by Urban and Gburek (1980) for a basin near Philadelphia. The results are summarized in Table 3.4. The basin was a permeable pavement where the only inflow was direct precipitation. Infiltration rate was assumed constant at the soil's hydraulic conductivity of 0.33 cm/h. In each time increment, storage = cumulative inflow - cumulative outflow = cumulative precipitation - cumulative infiltration. Storage increased rapidly during the early, intense hours of the storm. The time of maximum storage occurred at 12 hours. In more complex applications, other inflows and outflows can be added and time periods other than a single storm can be analyzed, but the principle of reservoir routing remains the same.

Long-term water balance of an infiltration basin

An infiltration basin's long-term water balance is a routing of flows over an average year. It sets the stage within which the basin responds to individual storm events and is critical to human amenity, landscape environment and the disposition of water resources. Early infiltration basin water balance models were those of Engstrand (1983) and Ferguson (1990).

A basin's long-term water balance is reflected in its hydroperiod or hydrologic regime. A dry basin empties entirely shortly after a design storm, and remains effectively dry during smaller background flows due to sufficiently rapid discharges through infiltration and evaporation. A wet basin has a stable level of standing water, established (where the basin is not in contact with the water table) by a combination of excess inflows and a surface outlet through which much of the inflowing water discharges. An ephemeral basin has standing water in at least some months; the level fluctuates from time to time as inflows accumulate and discharge by infiltration, evaporation and possibly surface overflow.

Table 3.5 and Figure 3.7 show the results of a basin routing using as inflow the direct runoff produced by the Chicago water balance introduced in Table 3.2 and Figure 3.4. This is a subsurface basin: void space is 38 percent; evapotranspiration and direct precipitation = 0. It can be seen that high runoff inflow in the spring and summer is discharged by more or less constant infiltration. However, despite the basin's large floor area, not all the inflow infiltrates immediately, due to the soil's low conductivity. Water is stored from month to month during the summer; this basin has an ephemeral regime. In the late summer, declining inflow allows the remaining stored water to infiltrate and ponding to cease. In the fall and winter all the low inflows are infiltrated as fast as they arrive.

Following is a description of routing equations in the order in which they are performed for each month (Ferguson, 1990). Table 3.5 is an example of the completed results. In these calculations, infiltration is shown only as an outflow, on the assumption that the basin is located above the water table. Hydraulic gradients are assumed constant at 1.0 through the floor and 0.7 through the sides. These calculations are easily set up in a spreadsheet; they can be combined in the same spreadsheet with the water balance of the drainage area. Calculations are iterative, December's accumulation carrying back to January and circular references cycling within each month.

Storage geometry

Geometry of water storage at the beginning of the month is calculated from the basin's given dimensions and the results of the previous month's water balance. If the basin is open, the volume of stored water is equal to the filled basin volume. If the basin has a stone fill, the volume of stored water is equal to the filled basin volume times the void space of the stone, commonly 0.38 to 0.40 for open-graded stone.

Stage = previous stage + previous Δstage;
Filled volume = Af (stage), where Af is floor area;
Hydraulic storage = (filled volume) (basin's void space);
Wetted side area = (stage) (basin's perimeter).

Table 3.5 Long-term water balance of a subsurface infiltration basin in Chicago. Runoff inflow is the direct runoff computed in Table 3.2, with drainage area of 1.25 ha. The basin is 1.0 m deep, 0.3 ha in area, and 230 m in perimeter length, with soil hydraulic conductivity of 0.05 cm/h, conductivity safety factor of 0.5, and side hydraulic gradient of 0.7.

	Jan	Feb	Mar	Apr	May	Jun	Jul	Aug	Sep	Oct	Nov	Dec
Inflow												
Direct runoff DRo (mm)	4	11	58	102	55	53	51	31	25	21	10	7
Storage geometry												
Stage (m)	0.00	0.00	0.00	0.08	0.60	0.63	0.66	0.66	0.45	0.21	0.00	0.00
Filled volume (m^3)	0	0	0	243	1941	2052	2151	2130	1470	684	0	0
Hydraulic storage (m^3)	0	0	0	92	738	780	817	809	559	260	0	0
Wetted side area (m^2)	0	0	0	6	52	55	57	39	18	0	0	0
Hydraulic flux												
Runoff inflow (m^3)	49	134	705	1239	668	644	620	377	304	255	122	85
Infiltration outflow (m^3)	49	134	612	594	626	607	628	627	603	515	122	85
Surface overflow (m^3)	0	0	0	0	0	0	0	0	0	0	0	0
ΔHydraulic storage (m^3)	0	0	92	645	42	37	-8	-251	-299	-260	0	0
Change in geometry												
ΔFilled volume (m^3)	0	0	243	1698	111	98	-21	-660	-786	-684	0	0
ΔStage (m)	0.00	0.00	0.08	0.52	0.03	0.03	-0.01	-0.20	-0.24	-0.21	0.00	0.00

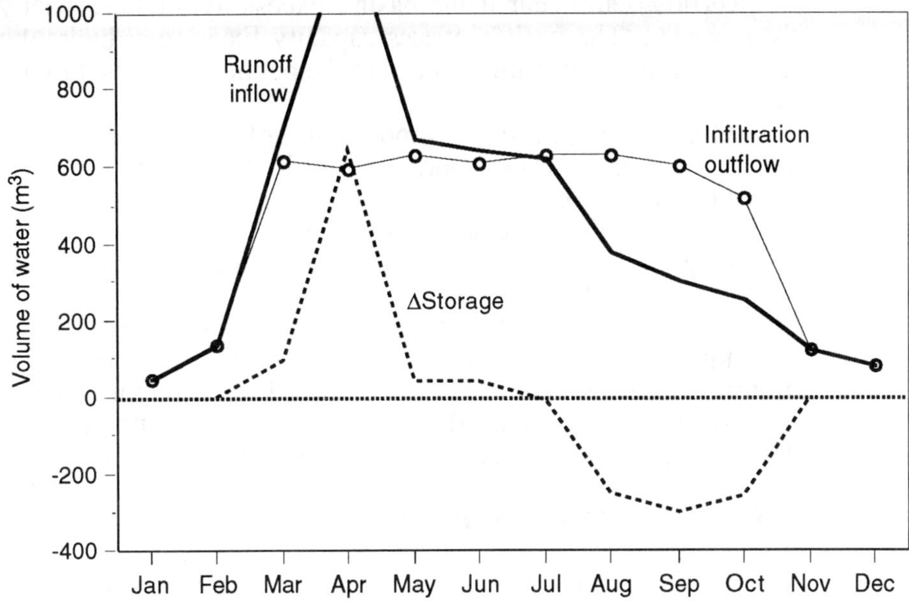

Figure 3.7 Long-term water balance of a hypothetical infiltration basin in the Chicago area (data from Table 3.5).

Flux during the month

Volume of inflow during a month is the sum of direct precipitation P, if any, and input from the drainage area:

Runoff inflow = (drainage area) (direct runoff depth);

Direct precipitation = P Af, if the basin is open;

Direct precipitation = 0, if the basin is closed.

Volume of outflow by evaporation and infiltration is based on the quantity of water in the basin available for outflow and on the basin's geometry. If the basin is open, evaporation can occur up to the limit of hydraulic storage + runoff inflow + precipitation inflow:

Evaporation = Et_o Af.

If the basin is subsurface,

Evaporation = 0.

Infiltration accounts for the remaining water, up to the limit of hydraulic storage + runoff inflow + direct precipitation - evaporation:

Infiltration = K Sf Af + 0.7 K (wetted side area).

Overflow can occur if the basin's excess hydraulic capacity is exceeded. The basin's excess capacity Ex is total hydraulic capacity minus hydraulic storage at the beginning of the month. Thus,

$$\text{Overflow} = \sum\text{inflow} - \text{evaporation} - \text{infiltration} - \text{Ex},$$
$$\text{if } \sum\text{inflow} - \text{evaporation} - \text{infiltration} > \text{Ex}.$$
$$\text{Overflow} = 0,$$
$$\text{if } \sum\text{inflow} - \text{evaporation} - \text{infiltration} \leq \text{Ex}.$$

Change in hydraulic storage is the difference between all inflows and all outflows:

$$\Delta\text{Storage} = \sum\text{inflow} - \text{evaporation} - \text{infiltration} - \text{overflow}.$$

Additional flows may occur in some basins, such as irrigation withdrawals or throughflows needed to maintain aeration for aquatic life. To take them into account appropriate additional terms need to be added to the flux equations.

Change in storage geometry

Change in geometry of water storage is based on the given change in hydraulic storage:

$$\Delta\text{Filled volume} = \Delta\text{hydraulic storage} / \text{void space}$$
$$\Delta\text{Stage} = \Delta\text{filled volume} / \text{horizontal area}$$

GROUND WATER MOUNDING

When infiltrating water reaches the water table, a mound of recharging water grows and spreads. The mound's shape and growth depend on the infiltration rate, the dimensions of the infiltration basin, and the hydraulic characteristics of the earth material. Mound height may need to be analyzed because of concerns of potential surface drainage problems, potential conflict with infiltration from the basin, or interference with pollutant removal processes in infiltrating water. Significant mounding may not occur where the water table is very deep.

Warner, Molden, Chehata and Sunada (1989) reviewed analytical solutions for the prediction of recharge mound height, including those by Glover (1960), Hantush (1967), Hunt (1971) and Rao and Sarma (1981). Huisman and Olsthoorn (1983) presented a detailed mathematical treatment of flow of recharged ground water in the vicinity of reclaiming wells.

Most analytical solutions solve some version of the linearized Boussinesq equation describing the flow of ground water in two dimensions from a rectangular basin (Warner et al.,

1989). Variables in the equation include aquifer transmissivity, water table elevation above the base of permeable material, and Cartesian coordinates with the center of the infiltration basin as origin. The hydraulic conductivity used in analyzing a ground water mound may be different from that used in analyzing infiltration from the surface. Infiltration is concerned primarily with vertical flow, while the ground water mound may be strongly influenced by horizontal flow. Layering of the soil may lead to different conductivities in the two directions.

Molden, Sunada and Warner (1984) developed a computer model using Glover's (1960) analytical solution. The model is capable of graphically displaying the rise and decline of a recharge mound for either an infinite homogeneous medium or a stream aquifer system. The model has been applied to evaluate infiltration strategies and to analyze the importance of possible variations in soil characteristics and boundary conditions.

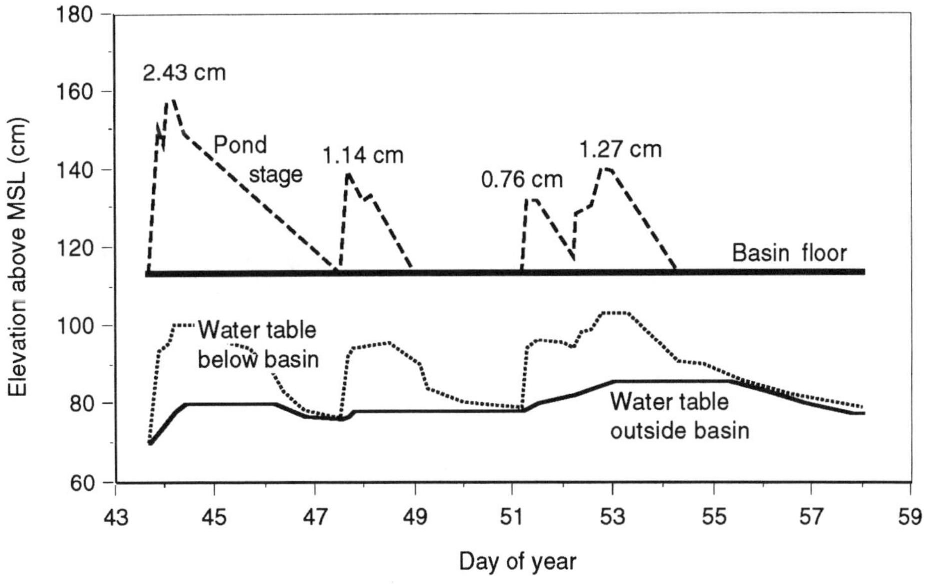

Figure 3.8 Ponding in the Surf City infiltration basin in North Carolina, and mounding in the underlying and nearby ground water (after Cheschair, Fipps and Skaggs, 1988). The amount of rain associated with each ponding event is shown above the ponding curve.

Figure 3.8 shows the water table below the Surf City basin in coastal North Carolina monitored by Cheschair, Fipps and Skaggs (1988). The water table was barely 30 cm below the basin floor. Percolation from the basin to the water table was rapid through the sandy soil. The mounding under the basin, fluctuating in close correspondence with ponding, contrasted with the mildly fluctuating water table outside the basin. In this sandy, conductive soil the mound dispersed and merged with the rest of the water table within a day or so after ponding and infiltration ceased.

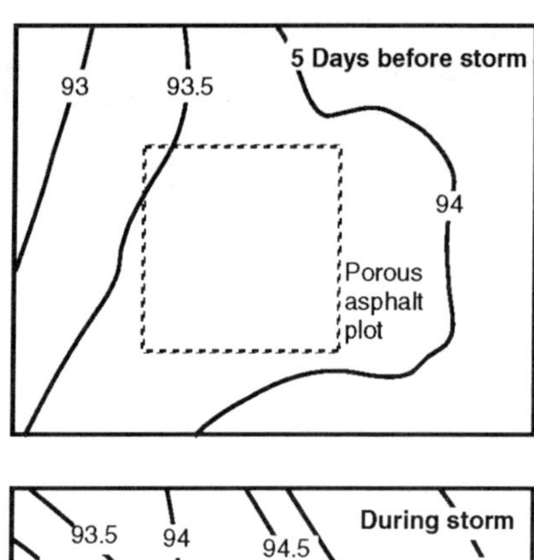

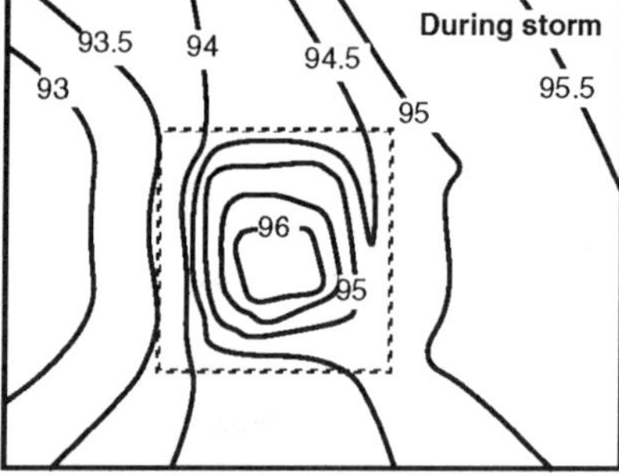

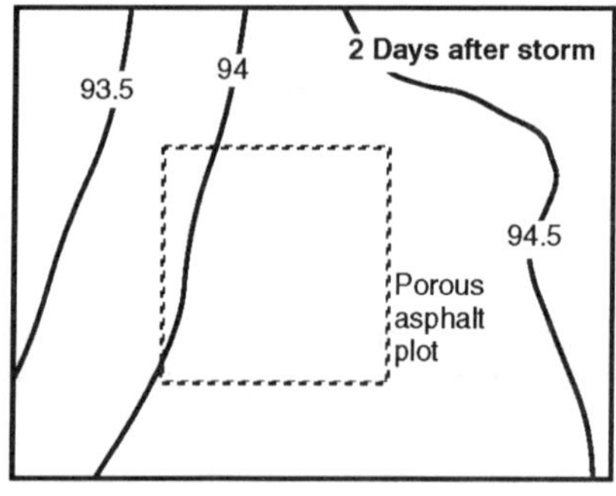

Figure 3.9 Ground water mounding under a porous pavement near Philadelphia during a 10 year storm (adapted from Gburek and Urban, 1980). Contours show elevation of the water table in meters above msl.

Near Philadelphia, Gburek and Urban (1980) monitored infiltration and recharge through the porous asphalt pavement shown in Figure 3.9. The pavement was surrounded by grass; its only inflow was direct precipitation. When percolate was generated, ground water mounding under the pavement was localized and temporary. The figure illustrates response to the 10-year storm of August 28, 1978 of 5.1 cm in 1.6 h. At the center of the plot ground water rose approximately 60 cm for every centimeter of rainfall. The mound dissipated within 3 to 5 days after its formation.

REFERENCES

Beven, Keith, and Peter Germann, 1982, Macropores and Water Flow in Soils, *Water Resources Research* vol. 18, no. 5, pages 1311-1325.

Bianchi, W.C., and Dean C. Muckel, 1970, *Ground-Water Recharge Hydrology*, ARS 41-161, Washington, D.C.: U.S. Agricultural Research Service.

Bouwer, H., 1966, Rapid Field Measurement of Air Entry Value and Hydraulic Conductivity of a Soil as Significant Parameters in Flow System Analysis, *Water Resources Research* vol. 2, no. 4, pages 729-738.

Brady, Nyle C., 1974, *The Nature and Properties of Soils*, New York: Macmillan.

Cheschair, G.M., Guy Fipps and R.W. Skaggs, 1988, Hydrology of Two Stormwater Infiltration Ponds on the North Carolina Barrier Islands, pages 313-319 of *Proceedings of the Symposium on Coastal Water Resources*, Bethesda: American Water Resources Association.

Chow, Ven Te, David R. Maidment and Larry W. Mays, 1988, *Applied Hydrology*, New York: McGraw-Hill, pages 39-40 and 99-123.

Dunne, Thomas, and Luna B. Leopold, 1978, *Water in Environmental Planning*, San Francisco: Freeman.

Engstrand, Daniel, 1983, *Retention Ponds, Design and Analysis of Ponds Without Outlets*, St. Paul: Minnesota Department of Transportation.

Ferguson, Bruce K., 1990, Role of the Long-Term Water Balance in Management of Stormwater Infiltration, *Journal of Environmental Management* vol. 30, pages 221-233.

Ferguson, Bruce K., and Thomas N. Debo, 1990, *On-Site Stormwater Infiltration, Applications for Landscape and Engineering*, New York: Van Nostrand Reinhold.

Ferguson, Bruce K., M. Morgan Ellington and P. Rexford Gonnsen, 1991, Evaluation and Control of the Long-term Water Balance on an Urban Development Site, pages 217-220 of *Proceedings of the 1991 Georgia Water Resources Conference*, Kathryn Hatcher, editor, Athens: University of Georgia Institute of Natural Resources.

Gale Research Company, (multiple dates), *The Weather Almanac*, Detroit: Gale.

Gale Research Company, (multiple dates), *Weather of U.S. Cities*, Detroit: Gale.

Gburek, William J., and James B. Urban, 1980, Storm Water Detention and Groundwater Recharge Using Porous Asphalt—Initial Results, pages 89-97 of *International Symposium on Urban Storm Runoff*, Lexington: University of Kentucky.

Glover, R.E., 1960, *Mathematical Derivations as Pertain to Groundwater Recharge*, Fort Collins, Colorado: U.S. Department of Agriculture, Agricultural Research Service.

Green, W.H., and G.A. Ampt, 1911, Studies on Soil Physics. I. The Flow of Air and Water through Soils, *Journal of Agricultural Science* vol. 4, pages 1-24.

Hantush, M.D., 1967, Growth and Decay of Groundwater-Mounds in Response to Uniform Percolation, *Water Resources Research* vol. 3, pages 227-234.

Heath, Ralph C., 1983, *Basic Ground-Water Hydrology*, Water-Supply Paper 2220, Washington, D.C.: U.S. Geological Survey.

Hershfield, David M., 1961, *Rainfall Frequency Atlas of the United States*, Technical Paper 40, Washington, D.C.: U.S. Department of Commerce, Weather Bureau.

Hillel, Daniel, 1980, *Applications of Soil Physics*, New York: Academic Press.

Horton, R.E., 1940, An Approach toward a Physical Interpretation of Infiltration Capacity, *Soil Science Society of America Proceedings* vol. 4, pages 399-417.

Huisman, L., and T.N. Olsthoorn, 1983, *Artificial Groundwater Recharge*, Boston: Pitman.

Hunt, B.W., 1971, Vertical Recharge of Unconfined Aquifers, *Journal of Hydraulics Division* (American Society of Civil Engineers), vol. 97, no. HY7, pages 1017-1030.

Jackura, Kenneth A., 1980, *Infiltration Drainage of Highway Surface Water*, FHWA/CA/TL-80/04, Washington, D.C.: Federal Highway Administration and Sacramento: California Department of Transportation.

Jensen, Marvin E., editor, 1990, *Consumptive Use of Water and Irrigation Water Requirements*, second edition, New York: American Society of Civil Engineers.

McCalla, T.M., 1945, The Influence of Microorganisms and Some Organic Substances on Water Percolation through a Layer of Peorian Loess, *Soil Science Society of America Proceedings* vol. 10, pages 175-179.

Miller, R.D., and Felix Richard, 1952, Hydraulic Gradients During Infiltration in Soils, *Soil Science Society of America Proceedings* vol. 16, no. 1, pages 33-38.

Molden, D.J., D.K. Sunada and J.W. Warner, 1984, Microcomputer Model of Artificial Recharge Using Glover's Solution, *Groundwater* vol. 22, no. 2, pages 73-79.

Morel-Seytoux, Hubert J., 1985, Conjunctive Use of Surface and Ground Waters, pages 35-68 of *Artificial Recharge of Groundwater*, Takashi Asano, editor, Boston: Butterworth.

O'Hare, Margaret P., Deborah M. Fairchild, Paris A. Hajali and Larry W. Canter, 1986, *Artificial Recharge of Ground Water, Status and Potential in the Contiguous United States*, Chelsea: Lewis.

Patterson, James C., 1976, Soil Compaction and Its Effects Upon Urban Vegetation, pages 91-102 of *Better Trees for Metropolitan Landscapes, Symposium Proceedings*, U.S. Forest Service General Technical Report NE-22.

Penman, H.L., 1948, Natural Evaporation from Open Water, Bare Soil and Grass, *Proceedings of the Royal Society* Series A vol. 193, pages 120-145.

Philip, J.R., 1957, The Theory of Infiltration. I. The Infiltration Equation and Its Solution, *Soil Science* vol. 83, no. 5, pages 345-357.

Prill, Robert C., and Donald B. Aronson, 1973, Flow Characteristics of a Subsurface-Controlled Recharge Basin on Long Island, New York, *U.S. Geological Survey Journal of Research* vol. 1, no. 6, pages 735-744.

Rao, N.H., and P.B.S. Sarma, 1981, Ground Water Recharge from Rectangular Areas, *Ground Water* vol. 19, no. 3, pages 270-274.

Rawls, W.J., D.L. Brakensiek and K.E. Saxton, 1982, Estimation of Soil Water Properties, *Transactions of the American Society of Agricultural Engineers* vol. 25, no. 5, pages 1316-1320 and 1328.

Rawls, W.J., K.L. Brakensiek and N. Miller, 1983, Green-Ampt Infiltration Parameters from Soils Data, *Journal of Hydraulic Engineering* vol. 109, no. 1, pages 62-70.

Rubin, Hillel, John P. Glass and Anthony A. Hunt, 1976, *Analysis of Storm Water Seepage Basins in Peninsular Florida*, Publication No. 39, Gainesville: University of Florida, Florida Water Resources Research Center.

Thornthwaite, C.W., and J.R. Mather, 1955, *The Water Balance*, Publications in Climatology No. 8, Centerton, New Jersey: Laboratory of Climatology.

Thornthwaite, C.W., and J.R. Mather, 1957, *Instructions and Tables for Computing Potential Evapotranspiration and the Water Balance*, Publications in Climatology vol. 10, no. 3, Centerton, New Jersey: Laboratory of Climatology.

United States National Oceanic and Atmospheric Administration (monthly and annual), *Climatological Data* (state), Asheville: National Climatic Data Center.

United States National Oceanic and Atmospheric Administration (monthly and annual), *Local Climatological Data* (station location), Asheville: National Climatic Data Center.

United States Soil Conservation Service, 1972, *National Engineering Handbook, Section 4, Hydrology*, Washington, D.C.: U.S. Soil Conservation Service.

United States Soil Conservation Service, 1986, *Urban Hydrology for Small Watersheds*, Technical Release 55, Second Edition, Washington, D.C.: Soil Conservation Service.

United States Weather Bureau, 1964, *Climatography of the United States No. 86*, Washington, D.C.: U.S. Department of Commerce.

Urban, James B., and William J. Gburek, 1980, Storm Water Detention and Groundwater Recharge Using Porous Asphalt—Experimental Site, pages 81-87 of *International Symposium on Urban Storm Runoff*, Lexington: University of Kentucky.

Warner, James W., David Molden, Mondher Chehata and Daniel K. Sunada, 1989, Mathematical Analysis of Artificial Recharge from Basins, *Water Resources Bulletin* vol. 25, no. 2, pages 401-411.

Woolhiser, David A., and D.E. Wallace, 1984, Mapping Average Daily Pan Evaporation, *Journal of Irrigation and Drainage Engineering* vol. 110, no. 2, pages 246-250.

4

Infiltration Hydrologic Design

The success of infiltration surfaces and basins in capturing, storing and infiltrating runoff at a given location depends on the relationship between their volumes and dimensions on the one hand, and the magnitude of precipitation and runoff inflow on the other. Each infiltration surface or basin must be sized for the flows anticipated to pass through it. Sizing must not result in underestimation of the required volume and floor area, so that excessive standing water and overflow will not ensue. Nor must it overestimate required capacity, so that land allocation and construction expense will not be wasted. Correct sizing involves the use of site-specific soil and watershed characteristics in calculations analogous to those for any other stormwater structure.

Both large storm flows and small background flows pass through every surface and basin. Infiltration basins, unlike other kinds of stormwater reservoirs, have no primary surface outlet. The entire volumes of small hydrographs, and the initial portions of large ones, are held without surface discharge. The relationships of these flows to proposed surfaces and basins can be analyzed using three kinds of models: design storm, long-term water balance, and continuous simulation.

Many basins have been sized by various empirical procedures, some inordinately conservative, due to concerns of safety, health, and liability (Weaver, 1969, p. 2), and some inordinately liberal due to lack of information. With continuing development of hydrologic science, experience with infiltration in the field, and computer simulation, design capabilities can be refined.

HYDRAULIC CONDUCTIVITY DATA

All infiltration design analyses require hydraulic conductivity of the soil at the basin site as one of the data inputs.

Reconnaissance data on subsurface conditions can come from published soil and geologic surveys, interpretation of aerial photographs and topographic maps (Way, 1978), and field examinations. Soil surveys published by the U.S. Soil Conservation Service are highly useful in characterizing general near-surface conditions in drainage areas and development sites, including soil texture, soil layering, and potential depths of water table and bedrock. In applying the results of regional surveys to specific basin locations published data must always be confirmed by direct field observation. Soil conditions existing before development may be greatly altered by grading and other construction activities, which can remove surface soil layers, add fill elsewhere, and compact all affected soil surfaces.

Site-wide surveys early in the planning process can be conducted to guide locations and general characteristics of infiltration basins and land uses. Seismic surveys can map hidden bedrock surfaces over large areas quickly. Piezometers (observation wells) can be installed for monitoring of fluctuations in water table elevations. Appropriately planned drilling can log subsurface textures and water table depths, and provide samples for laboratory testing. Drilling with sufficient frequency can disclose horizontal variations in soils. Depending on the type of environment and the size of proposed basins, the depth of drilling might be to the water table, to the bedrock surface, or at least to 1.5 to 5 m below proposed basin floors. If borings are conducted for other aspects of a development, such as structural design of building foundations, and the boring locations are at appropriate locations and adequate in number, their results can be interpreted for design of infiltration surfaces and basins as well.

Given reliable soil data, hydraulic conductivity can be estimated from soil texture. Rawls, Brakensiek and Saxton (1982) gave the average saturated hydraulic conductivities for soil textures listed in Table 3.1. The Maryland Water Resources Administration (1984) adopted these values in its statewide infiltration design standards on the premise that soil texture can be identified easily and conclusively. This allows design and construction to proceed quickly and economically on a known basis, whereas further site-specific testing of soils could result in variable and arguable conductivity values despite the time and expense required to obtain them. Wherever conductivity is based on texture, texture must be confirmed by direct field observation.

However, soil texture may provide only the roughest of guides to conductivity in the field. Where a crust develops on a soil surface after construction, the crust can be less conductive than any soil layer that existed before construction, and can become the most limiting layer in the soil profile for infiltration. On the other hand, where vegetative growth and other processes create macropores, the surface soil can be more conductive than texture alone would have predicted. Thus structure—the aggregation of soil into units larger than individual soil particles—can be more important than texture to hydraulic conductivity. Structure can be altered, advertently or inadvertently, after construction by vegetation, sedimentation, soil development and chemical additives. At this time not enough is known about soil structural development and its effect on conductivity to use it in quantitative design of proposed surfaces and basins. Until we gain this knowledge, guides such as Table 3.1 will be necessary in design. In some areas, soil texture is used to form a first approximation of conductivity for use in early planning, after which conductivity tests are performed at the exact locations of proposed basins. In any case, following design and construction, basins should be monitored for their hydraulic performance and general development. The management of soil structure as part of basin maintenance is described in Chapter 6.

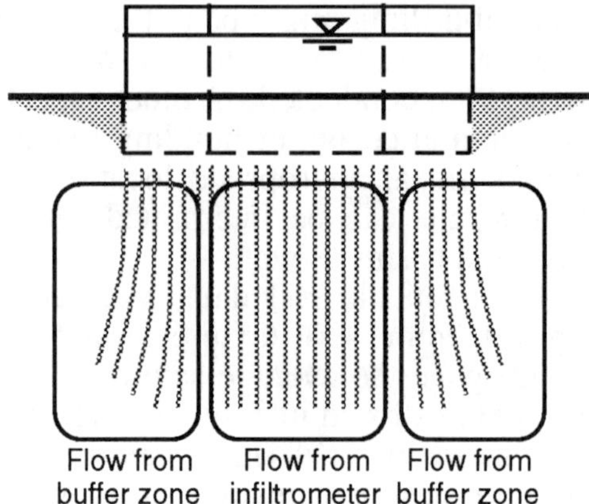

Figure 4.1 Double-ring infiltrometer.

Testing

Field and laboratory tests for saturated hydraulic conductivity were described by Amoozegar and Warrick (1986), Dunne and Leopold (1978) and Weaver (1969, p. 32-40). Some have been standardized by the American Society for Testing and Materials. All tests must be conducted under known hydraulic head (ponding depth) so that Darcy's law can be solved for K. They must be continued long enough—in some cases more than a week—to reach a stable final K.

In the field, tests on in situ soil can be carried out after excavating below the proposed basin floor (Paul Thiel Associates, 1980). The number of test sites depends on the extent of the contemplated infiltration basin and on variability of soils in the area. Any small soil area that is isolated for the measurement of water intake is called an infiltrometer (Bianchi and Muckel, 1970, p. 30).

A ring or flood infiltrometer contains ponded water and measures the amount entering the soil. Its applications, procedures and sources of error were reviewed by Bouwer (1986). A small model basin is constructed by pushing a section of pipe or a bottomless box into the soil, or by constructing an impermeable dike around a soil area. The basin is filled to a given depth (head) with water and maintained at that depth while infiltration continues. The infiltration rate is equal to the rate of replacement of water (subject to evaporation, leakage and other losses).

A double-ring or concentric-ring infiltrometer is widely used. It attempts to measure only vertical infiltration by eliminating the effects of lateral spreading. As shown in Figure 4.1, an outer concentric ring or box surrounds the infiltrometer and is kept filled simultaneously with the infiltrometer. The area between the inner and outer rings serves as a buffer, confining infiltration from the inner basin to vertical movement. Infiltration rate is measured only in the inner basin, giving an estimate of true vertical rate. Various heights, widths, numbers of rings and measuring methods have been used, all having their benefits and deficiencies.

Prill and Aronson (1978) used a ponding test on a basin on Long Island with very permeable soil. They diked test plots small enough that the infiltrating volume would not exceed the available water supply. They placed piezometers at soil depths expected to be under positive pressure during a test, and tensiometers at depths expected to be under negative pressure. Following ponding tests, hydraulic conductivity in each soil zone was derived from Darcy's law, using measured flow rate and change in head between upper and lower boundaries of the zone.

A sprinkling infiltrometer applies a certain amount of artificial rainfall to a plot positioned on an incline. At the bottom of the incline a trough is used to collect and measure the runoff.

Specific types of test are designed for soils where the water table is shallow (Amoozegar and Warrick, 1986). The auger-hole method is widely used: a cavity is dug below the water table, with minimum disturbance of the soil; the water in the hole is pumped out, and the rate of rise of the water level measured. The piezometer method is similar in principle, but a pipe is installed into the auger hole, so that water can enter the pipe only from the bottom. The double-tube method is like the double-ring infiltrometer, with two concentric cylinders pressed into the earth and a constant water level maintained in both cylinders. After saturation of the soil, the flux from the inner tube into the soil is measured; the flux from the outer tube confines movement from the inner tube to the vertical dimension.

Laboratory measurements on soil samples may be made if in situ measurements are not possible or not desired. Soil cores are obtained by pressing a slender cylinder into the surface and extracting a core of soil for study. Klute and Dirksen (1986) reviewed sampling, apparatus and procedures for laboratory measurement. Laboratory tests are based on imposing a hydraulic head on a soil column of uniform cross-sectional area, measuring the resulting flux of water, and solving Darcy's law for the hydraulic conductivity. Variations include constant-head and falling-head methods. Laboratory tests are considered only an approximation: soil structure and porosity are easily disturbed during removal of a sample from the ground and setting it up for testing in a lab.

Assigning hydraulic conductivity

The hydraulic conductivity assigned for design of an infiltration basin should be a typical value which will apply for the life of the basin.

The assigned conductivity should be based on the most limiting of the soil layers at or below the basin floor. The limiting layer could be a dense, fine-textured crust that develops after construction is completed and the basin is in operation. In subsurface basins incorporating a filter layer or fabric, the filter layer may be the most limiting layer in the soil profile. In a large basin in contact with more than one soil type, the assigned conductivity can be an average weighted on the basis of the basin area underlain by each type.

The magnitude of K for a given material involves great uncertainty. Neither on-site testing nor indirect indicators such as soil texture give completely consistent or reliable results. Testing of an existing soil before construction does not take into account alteration of conductivity during and after construction by compaction, sediment accumulation and soil development. It is essential that the conductivity used in design be conservative (low) enough that future infiltration will not be still lower.

A way to take uncertainty into account in design is with a safety factor. Paul Thiel Associates (1980, p. 180) have applied in practice a safety factor Sf of 0.5 to saturated hydraulic conductivity; thus the K they assume for design is effectively half that yielded by tests or other indicators. The same value of Sf is specified in a manual by the Washington State Department of Ecology (1992, p. III-3-16). The value of 0.5 is coincidentally equal to that found necessary by Bouwer (1966) to take into account the trapping of air in subsurface voids during sorption. In order to take into account other factors such as compaction, sedimentation and microbial growth, Sf may need to be lower than 0.5. A basin's floor is particularly subject to compaction from traffic, crusting, and accumulation of sediment, compared with steep basin sides.

Live vegetation can maintain high surface conductivities in aerobic soils by maintaining soil structure and porosity, even through slowly accumulating layers of sediment. However, a vegetated surface does not necessarily allow a high value of Sf in design. Where mesic vegetation is to survive, the assigned conductivity must be reduced to prevent the vegetation from suffering from a saturated, poorly aerated root zone receiving inflow at the maximum rate determined by saturated hydraulic conductivity.

For a subsurface basin, a conservative (low) value of Sf addresses the potential obstruction of a portion of the soil surface by particles of stone fill.

HYDRAULIC GRADIENT DATA

Simple design methods assume a constant conservative vertical hydraulic gradient of 1.0, yielding conservative estimates of low cumulative infiltration and long ponding time. On basin sides hydraulic gradient is less than on the floor due to the lateral direction of infiltration. For example, the gradient is 0.7 if the direction of flow is 45° from the vertical.

Taking into account varying gradient using the Green-Ampt model requires additional computational effort. When this effort is borne, it is rewarded by a more valid estimate of all factors, shorter estimated ponding time and smaller required basin.

DESIGN STORM ANALYSIS

A design storm is a brief, isolated runoff event such as that with a 2- to 10-year recurrence interval, or a fixed flow quantity such as the first 2.5 cm of runoff from every storm. For a given recurrence interval, rainfall amount and intensity vary from place to place. Control of runoff during a design storm is important to downstream flooding, erosion and water quality.

An infiltration basin sized for a large, infrequent storm controls all storms up to the design storm effectively, but requires expense in proportion to its great size. Sizing it for a smaller, more frequent storm reduces expense, but produces only partial control of larger storms.

Many local agencies specify design-storm conditions to be used in their areas. For example, some require a single 50-year storm or the first 2.5 cm of runoff from every storm, others require two 10-year storms separated by a given time period, and still others require combinations of 2 -year, 5-year, 10-year and other storm frequencies. Where design criteria are specified only for conveyance (pipes and channels) or detention, the same conditions can justifiably be applied to infiltration as well.

Hydraulic capacity

Hydraulic capacity is the volume of a surface basin, or the volume of void space in a subsurface basin. A proposed basin's total volume can be found from a grading plan using any of the methods used in earthwork calculations: contour planes, average end area or grid. The void volume in a subsurface basin is the total basin volume times the void space in the fill. The void space is typically 0.38 to 0.40 for open-graded crushed stone.

In a subsurface basin, hydraulic capacity may be layered, especially in porous pavements where different sizes of aggregate are used in the various pavement layers (Paine, 1990). Figure 4.2 shows an analysis of a pavement system for capacity. Each successive layer has a unique storage capacity based on its porosity and depth; the depth times the porosity is the equivalent depth of water that can be held. Adding the capacities of the various layers gives the capacity of the entire pavement or basin. Where appropriate, temporary storage over the pavement, created by a curb or dike around the pavement storage area, can also be counted in capacity for extremely large storm events.

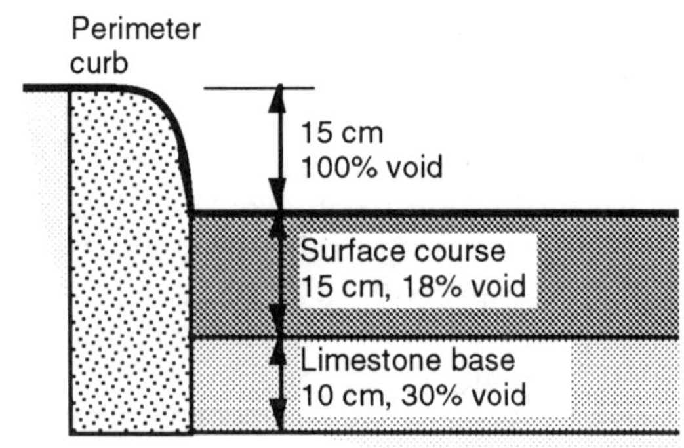

Figure 4.2 Analysis of a layered pavement system for hydraulic storage (after Paine, 1990).

A basin could be sized to fulfill a variety of volume objectives:

• To infiltrate the entire inflow volume (Q_{vol}), allowing no surface discharge from the drainage area. The cost of a large basin is offset by the elimination of pipes and other primary conveyances immediately downstream. This type of design is required on Long Island, where an early motive for infiltration was reduction in downstream conveyances. It is highly effective at controlling downstream floods and erosion, capturing pollutants and returning a large portion of annual runoff to the ground water and base flow.

• To infiltrate the increase in volume of storm flow (ΔQ_{vol} or ΔDRo) compared to predevelopment levels, thus keeping surface discharge volume the same as before development. The fraction of the runoff flow in excess of the capacity overflows on the ground surface during the design storm. Gutters, swales, channels or pipes may be necessary to carry the overflow to appropriate discharge points during the design event. In Maryland, infiltration basins must hold the increase in design-storm volume attributable to development of the drainage area (Maryland Water Resources Administration, 1984). This requires estimating the volume of runoff before and after development using the SCS method; the difference is the required infiltration volume. This type of design is very equitable: it allocates infiltration cost among individual developers in proportion to the runoff impacts they create on their own sites. It can be effective at flood and erosion control, water quality improvement, and maintenance of ground water and stream base flow.

• To infiltrate a fixed volume of runoff from every storm event. In Florida infiltration basins are sized to hold only the first flush of 1.25, 2.5, or 5 cm of runoff from every storm (Wanielista and others, 1981). Finding a required basin capacity requires only measuring the size of the drainage area and multiplying by the required runoff depth. This type of design tends to be applied where the principal motivation for infiltration is water quality improvement. It does not necessarily capture a large proportion of the volumes of large storm flows for flood and erosion control.

• To reduce overflow to a given rate, for instance the peak rate of discharge (Q_p) from the drainage area before development. This kind of objective is embodied in many existing stormwater ordinances in the United States, although detention rather than infiltration has been the principal means of fulfilling it in the past. The motivation for this kind of design is downstream flood control. As shown in Figure 3.6, any amount of infiltration up to $0.375Q_{vol}$ produces no reduction in peak rate of discharge (overflow) compared with the flow arriving from the drainage area; the later flows, including the peak, pass over the basin without modification. Basins holding greater than $0.375Q_{vol}$ reduce peak flow at the point of discharge, because they capture the water traveling at the peak rate. The degree of peak flow reduction for any amount of infiltration greater than $0.375Q_{vol}$ can be read from the graph, or derived by modeling flow volumes and rates during a storm event.

Sizing a basin for the full runoff volume or for a fixed portion of the volume during a given storm event is a simple procedure. The calculation does not take into account infiltration outflow during the storm, so it yields the largest volume that the basin might require at any time during the storm event. In some areas basin capacity is derived directly by locally developed empirical formulas analogous to the rational formula (Hannon, 1980, p. 138-139). In San Joaquin County, California, basin volume is derived by V = CAR, where V is the required volume of the basin, C is the runoff coefficient in the rational formula, A is drainage area, and R is rainfall depth during the 10-year storm. On Long Island similar formulas are used by local agencies.

Taking into account infiltration during a storm filling period reduces the required storage volume and floor area, but it is a more complicated routing procedure and its validity is dependent on very reliable assignments of hydraulic conductivity and the inflow hydrograph. Using this procedure, a basin's capacity must equal the peak difference between the mass curves for inflow and infiltration (Weaver, 1969, p. 2). In Table 3.4, the basin volume was intended to hold the volume of runoff from the 25-year storm, less infiltration during the event. Hence the required design capacity is 8.89 cm, the maximum found during the design event.

Hannon (1980, p. 139-146) and Weaver (1969, p. 28-29) described use of the rational formula by the New York State Department of Transportation on Long Island to derive a mass curve of cumulative inflow. A mass curve of rainfall is constructed from an appropriate rainfall intensity-duration-frequency curve. It is multiplied by the weighted runoff coefficient (c) for the drainage basin to produce the mass curve of runoff, with upward adjustment as a safety factor. A mass infiltration curve is constructed by assuming a constant infiltration rate in a unit basin floor area. The greatest difference between inflow and outflow at any one time is the required basin volume. A combination of depth and floor area is selected to store that volume.

A more versatile approach is taken in Miami and northern California (Hannon, 1980, p. 148-155). In these locales runoff coefficients are applied to the 3 year storm to construct mass inflow curves. Mass outflow curves are constructed at the minimum allowable infiltration rate and adjusted downward by a safety factor. The difference is either required storage or the overflow that must be handled by further downstream infiltration.

Thelen and Howe (1978, p. 49-50) described a storm routing that explicitly takes into account an allowable release rate. The cumulative inflow is found as the sum of inflowing runoff and direct rainfall. The cumulative outflow is the sum of allowable overflow from the basin, for instance a constant maximum permissible rate, and infiltration. The greatest difference between cumulative inflow and cumulative outflow at any time is the required basin capacity.

Debo (1994) has constructed a computer model for sizing basins for design storms. The model estimates storm hydrographs before and after development using the SCS (*TP-149*, U.S. Soil Conservation Service, 1968) method. It uses the results to derive volumes, depths and other dimensions of surface and surbsurface basins based on given constraints such as soil conductivity and allowable drawdown time. The vocabulary of basin types is similar to that of the Maryland Water Resources Administration (1984).

Accommodating throughflow rates

Ponding time during a design storm is a concern where hydraulic capacity, multiple use or environmental conditions must be restored quickly following the storm.

The allowable drawdown time should be appropriate to a given storm frequency. Very short ponding times should be expected only during small, frequent storms that generate little runoff volume. Relatively long ponding times should be expected during large, infrequent storms. A range of frequencies could be evaluated, with longer ponding times allowed for larger storms.

The allowable drawdown time should also be appropriate to the purposes of a given infiltration basin and its environmental design. A maximum ponding time of only one day might be appropriate during a substantial storm where a basin is used for competitive sports or other rigidly scheduled activities. Longer times of several days or more might be allowed where the schedule of use is not so rigid. Allowable ponding time may also be limited by the need to protect the health of inundated vegetation, to maintain aerobic soil for water quality treatment, or to prevent mosquito breeding.

One way to estimate ponding time is to assume that infiltration occurs only through a basin's floor and not the sides. Under these conditions, given a captured volume of water and a constant infiltration rate equal to K, ponding time T_p is controlled by depth. Basin depth is derived to produce the required ponding time:

$$\text{Maximum permissible depth} = KSfT_p/V_d.$$

where

Sf = infiltration rate safety factor;

V_d = basin void space.

The less the depth, the shorter the ponding time. This derivation is embodied in Maryland's (1984) statewide infiltration standards.

For a given volume to be infiltrated in the required time, the basin's floor area must be large enough to spread out the volume to the required depth. Based on the equality depth = volume/area,

$$\text{Minimum floor area} = \text{volume/maximum depth}.$$

If some infiltration is assumed to occur through a basin's sides, then increasing the basin's perimeter by laying out long, narrow shapes or complex fingers and peninsulas that reach into available interstices in the site increases infiltration while water is ponded, and decreases ponding time. Modeling infiltration through the basin sides requires a reservoir routing procedure for the design storm, taking into account the changing values of depth and wetted side area during infiltration.

In Sweden, infiltration is assumed to occur only through a basin's sides (Stahre and Urbonas, 1989). The basin's floor is assumed impervious due to clogging by sedimentation, leaving only the sides for unclogged infiltration.

Where perforated pipe is used to distribute inflow through a subsurface basin, the number and size of the perforations must be sufficient to pass the expected flow rate. Duchene and McBean (1992) in the laboratory measured steady exfiltration from a level perforated pipe. The pipe was surrounded by 5-cm gravel, thereby taking into account the possibility that some of the perforations could be partly or fully blocked by gravel particles. They found that, for a 300-mm-diameter smooth-walled pipe with 12.7-mm-diameter perforations, exfiltration could be described by the orifice equation, $Q = CA(2gH)^2$, where Q = discharge (m^3/s), C is a coefficient equal to 0.65, A is the total area of the submerged orifices (m^2), g is acceleration due to gravity (9.82 m/s^2), and H is the head to the center of the orifice (m).

Despite such refinements, storm-based design criteria do not determine a basin's ongoing hydrologic character or the disposition of a significant part of the total water resources of a drainage area. In the New York City area, where the annual precipitation averages about 110 cm, a 10-year storm is only 12 cm in 24 hours. The storm lasts only a minute fraction of one percent of the elapsed time during its recurrence interval. Eleven meters of rain go by year by year, while stormwater basins wait for the mere 12 cm for which they were designed. A broad interest in water resources, the environment, and the functioning of infiltration basins demands a more encompassing design approach.

WATER BALANCE ANALYSIS

The long-term water balance is a way to take into account the low, continuous background flows that pass through every infiltration surface and basin. It supplements the analysis of isolated pulses of runoff in design storms. The long-term hydrologic characteristics relevant to a basin's multiple purposes—aesthetics, safety, wildlife, recreation—are a result of flows that continue during the extensive periods between design storms and of the basin's resulting hydrologic regime—dry, wet or ephemeral. Amounts of ground water recharge and base flow that a basin induces are represented better by total annual flows than by individual storms. Taking into account the volume of cumulative background flows in design assures that a basin's capacity is not preempted by standing water before a design storm begins.

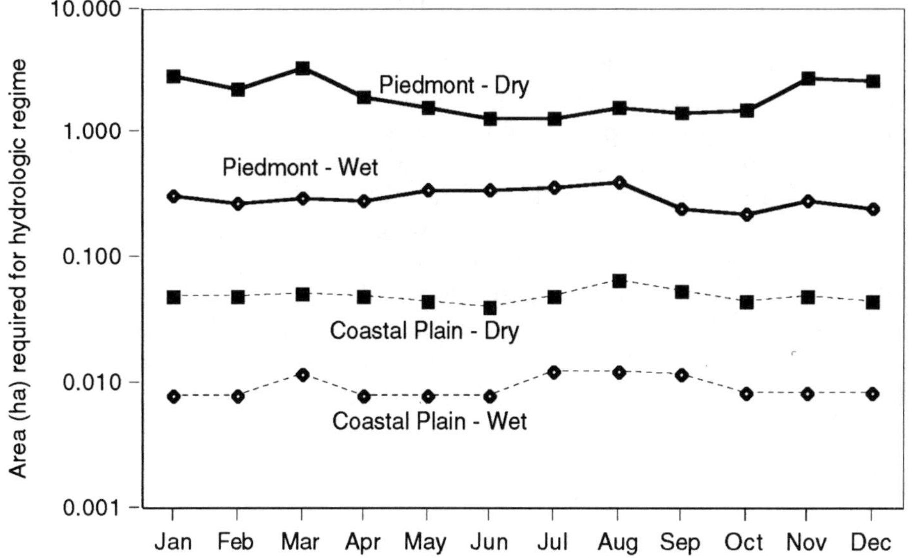

Figure 4.3 Floor areas required to establish dry and wet regimes in infiltration basins with hypothetical 4-ha drainage areas in the Philadelphia area (data from Ferguson, 1990a). The Piedmont site has high runoff and low soil hydraulic conductivity; the Coastal Plain site has low runoff and high soil hydraulic conductivity.

The water balance, like other hydrologic models, is a simulation model. Given the rain that will fall on a certain drainage area and infiltration basin, the water balance estimates what will happen. Using it in basin design is a trial-and-error process. Trial basin dimensions are chosen and then the hydrologic functioning of the basin is evaluated by applying the model.

To illustrate the water balance's implications for site-specific design choices, Ferguson (1990a) applied monthly basin routing to hypothetical 4 ha multi-family residential sites in the Piedmont and Coastal Plain in the Philadelphia area. Although water balance modeling has been refined since that study was done, the relative differences between the sites and the alternatives in site layout that they bring to mind are instructive. Figure 4.3 shows the different floor areas required to establish dry and wet regimes as the water balance fluctuated over the seasons.

Table 4.1 Values of conductivity safety factor Sf used to protect terrestrial vegetation in specific soil and slope conditions (Wade Nutter, University of Georgia School of Forest Resources, personal communication, 1985).

Depth of most limiting soil horizon	Sf
Sloping site with free drainage	
0.6 m or less	0.10-0.15
Deeper than 0.6 m	0.20-0.25
Level site with little free lateral drainage	
0.6 m or less	0.05-0.07
Deeper than 0.6 m	0.10-0.12

Dry regime

To establish a dry regime, a basin's floor area is made sufficiently large to dispose of all inflows by infiltration and evaporation in the month when they occur, such that monthly stage never exceeds zero. A basin large enough to dispose of inflows during the wettest month of an average year can stay dry the rest of the year as well. Dry basins have the cost advantage that the entire volume is reserved for the design storm; none of the constructed capacity is preempted by standing water before the design storm begins. However, a large land area must be allocated to the floor to establish the dry regime. A dry regime permits multiple human use such as playing fields, and helps maintain vegetation and soil aeration for organic matter decomposition and for maintenance of soil structure and porosity.

In order not to drown the roots of mesic plants, the allowable percolation must be reduced below that possible under saturated hydraulic conductivity only. The safety factor Sf on conductivity might be 0.04 to 0.10, with the lower values being used where soil conditions are variable or poorly defined (U.S. Environmental Protection Agency, 1981). Some researchers allow slightly more liberal safety factors under certain conditions. For example, Wade Nutter, a forest hydrologist, uses the factors listed in Table 4.1.

The floor of a dry basin must be constructed level in order to assure that all flows are distributed uniformly, taking advantage of the full floor area. Over time, an alluvial fan of accumulating sediment could cause low-flow runoff to concentrate upon a small portion of the basin floor, where outflows cannot operate at their full rates. It may be prudent to reserve a portion of the floor near the inlet for sediment storage, in addition to the minimum size required to satisfy the water balance.

On Ferguson's (1990a) Coastal Plain site, the soil's high hydraulic conductivity and low runoff produced a dry regime on a 0.06-ha basin floor, only 1.6 percent of the catchment. Such a small area is very likely to be acceptable from a land allocation viewpoint. It would be easy to allocate more of the site area to the basin, reducing ponding depth and time, without compromising the site's basic urban land use. For instance, if a limit on ponding time of 2 days is desired for the 10-year storm, the basin would have to be 0.20 ha in area.

The Piedmont site's low soil conductivity and high runoff required a basin floor area of 3.7 ha to stay dry year-round. This was more than 50 times the area required on the Coastal Plain site, and nearly as big as the basin's 4-ha catchment. Thus a dry single-purpose surface basin was less feasible on the Piedmont than on the Coastal Plain because of the Piedmont soil's lower conductivity and higher runoff potential. The limiting month was March, the wettest month of the year. This basin's lavish requirement for space stimulated a search for alternatives to an open, dry, single-purpose basin. Some alternatives are:

- to take up as much of the basin area as possible in subsurface basins, reclaiming the use of part of the land surface;
- to fit multiple use such as recreation or created wetlands into a surface basin, retrieving some of the land's economic or environmental value;
- to make the basin smaller in area and to treat the landscape design appropriately for the ephemeral or perennial pool that could then be expected to develop; and
- to reduce the inflows by converting upstream pavements to porous materials.

In choosing among such alternatives, the costs of the alternative treatments have to be weighed against the development's land use objectives.

Wet regime

To establish a wet regime, a basin's floor area is made sufficiently small not to dispose of all inflows by infiltration and evaporation, in most cases bringing about surface discharge of excess water. A basin small enough to have standing water during the driest month of an average year can remain wet the rest of the year as well. For wet basins with side slopes of less than about 20 percent (5H:1V), the entire basin area can be treated as a level floor with vertical infiltration, without a significant loss of accuracy (Ferguson, 1990a).

In order to assure a wet regime, soil hydraulic conductivity must not be underestimated. The conductivity safety factor Sf might be raised to 1.0 or higher.

In Ferguson's (1990a) Philadelphia illustrations, the Piedmont site could support a wet basin of 0.27 ha in area year-round. A pool of such a size could be used as a visible and significant human amenity on a 4-ha urban site. It was supportable without an artificial lining due to the naturally low conductivity of the Piedmont's soil.

The Coastal Plain site could support an unsealed wet basin year-round of only 0.01 ha due to rapid losses into the sandy soil. Such a small pool is unlikely to be a significant amenity on a 4-ha site. A larger unsealed pond could be supported if water supply from wells or city lines is sufficient to supplement runoff in dry months; an appropriate inflow term could be added into the routing to investigate specific quantities. An alternative would be to seal the bottom of the basin, but this would reduce or eradicate the basin's infiltration function.

Ephemeral regime

An ephemeral regime uses basin walls to infiltrate water as well as the floor. This provides a buffer against the possible development of a low infiltration rate on the basin floor resulting from crusting, compaction or sediment accumulation.

In water balance calculations for an ephemeral basin in an average year, in which December's results cycle iteratively back to the beginning of January, the results are frequently unstable—the solution fails to converge on a single set of values—due to the feedback between the amount of water stored and the rate of infiltration through large side areas.

The hydrologic performance of basins with ephemeral regimes is particularly amenable to analysis by the Minnesota method.

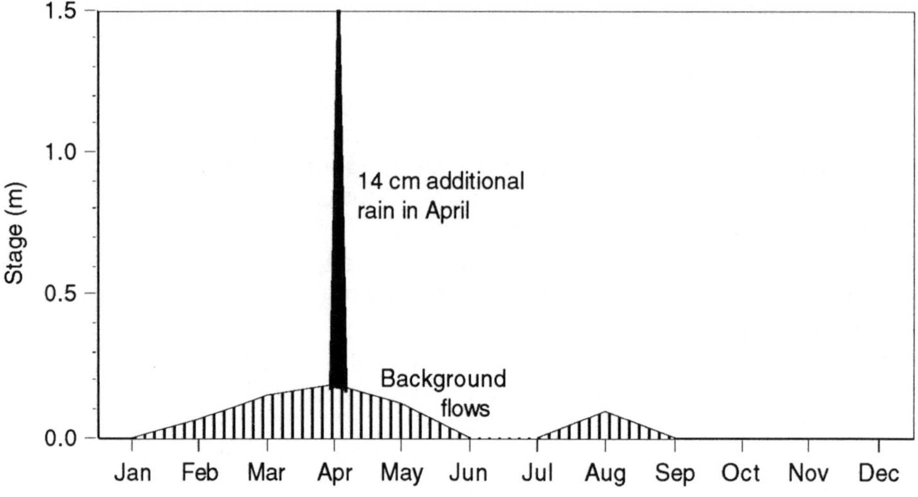

Figure 4.4 Application of the Minnesota method to an ephemeral basin in Athens, Georgia. The basin is 1.5 m deep.

MINNESOTA METHOD

Standing water resulting from the long-term water balance preempts the capacity of a basin, reducing its capability to contain the volume of a design storm. Thus the long-term water balance and the design storm work together to determine sufficiency of basin capacity.

The Minnesota Department of Transportation (Engstrand, 1983) used a water balance to investigate the possibility of disposing of highway runoff in existing glacial kettle holes. A water balance for each month, including inflow of proposed highway runoff, was used to estimate seasonal water level fluctuations. The 100-year storm volume was then superimposed on the highest monthly water level. If a kettle hole did not overflow under these conditions, it was considered capable of receiving the highway runoff.

Ferguson (1990b) adapted the Minnesota approach to the design of proposed basins. Trial basin dimensions are chosen, sufficient at least for the required volume of the design storm and taking into account the volume of stone fill. Average monthly flows are routed through the trial basin. The monthly calculations are reviewed to find the month with the smallest reserve (unfilled) volume. If the reserve is smaller than that needed to capture the design storm's runoff, the dimensions of the basin are adjusted and the calculations repeated until a satisfactory basin is found.

Figure 4.4 shows an application of the Minnesota method to a stone-filled basin in Athens, Georgia. The hatched area represents the volume of water stored in the basin from month to month. The white area represents unused hydraulic capacity held in reserve for a design storm. The dark spike shows the effect of an occurrence of the 10-year storm in April, when the average background storage is greatest. In this event the basin is filled to capacity. Infiltration from the basin restores the basin's capacity within a few days after the storm.

Ferguson (1990b) used the Minnesota approach to model the behavior of hypothetical basins in the Atlanta area with various capacities and dimensions. Large overflows during the 10 year storm occurred in small basins such as those with capacity of 2.5 cm of runoff. This resulted partly from the basins' small capacities, which are typically intended for first flush quality treatment rather than for flood control. However, overflows were greater than conventionally anticipated, because the basins' limited capacities tended to be at least partly preempted by standing water many months during the year. Overflows occurred even in some basins with capacities equal to the volume of the design storm, if their floor areas were small enough that they were partly filled by background flows before the design storm occurred.

Following simulation of a basin's behavior with the Minnesota method, the basin's performance can be evaluated using the following criteria:

• Proportion of design-storm runoff volume held results from the unfilled volume during the wettest month in the long-term water balance. The unfilled (reserve) capacity must be big enough in the month with the highest monthly stage. If it is not, depth and thus capacity can be increased, or infiltration can be hastened by enlarging the basin's area or perimeter or finding a way to make the bottom more permeable such as by puncturing an impeding layer with a supplemental well.

• Standing water prior to an occurrence of the design storm preempts some of the capacity of the basin, making the total required volume (and therefore, in many cases, construction cost) greater than it would be for a dry basin. Thus if volume and cost are to be minimized it is imperative to make a basin's floor large enough to infiltrate all background flows without the accumulation of standing water. Standing water can be reduced by enlarging a basin's floor area, causing background flows to infiltrate more quickly. Infiltration can be further hastened by enlarging a basin's perimeter, increasing infiltration through the sides. However, infiltration through the sides is significant only as long as the water is ponded to a significant depth; perimeter length matters little when the water is at a low stage and is infiltrating mostly through the floor. Complete elimination of standing water requires an adequately large floor area, almost irrespective of perimeter length.

• Unused volume at the top of the basin, above both maximum standing water and the required design-storm volume, indicates oversizing and possibly unnecessary construction expense. To reduce this figure, basin depth can be reduced.

• Overflow during the design storm is volume in excess of that captured by a basin. Overflow runoff must be conveyed safely across the site to a discharge point, or eliminated by enlarging the basin volume to increase capacity or by enlarging the floor area to reduce standing water.

• Overflow during a wet month, before the design storm occurs, is a loss of background flows for ground water recharge and base flow support. It indicates a lack of reserve storage in the basin to capture the design storm should it occur. Perimeter length or floor area can be increased to hasten infiltration, or depth can be increased to enlarge basin capacity.

A basin is selected that has capacity and dimensions adequate for both the accumulation of background flows and the required design storm. Where the floor areas required for ponding time and long-term regime differ, the larger area is selected.

The Minnesota method was used to set design standards at Coggins Park, an industrial park in Athens, Georgia. Builders on individual lots were to infiltrate their runoff on-lot according to criteria set by the park's developer. Basins were to catch the runoff from at least 90 percent of the area of each lot and have floor area sufficient to assure a dry regime. Based on the long-term water balance in the park's fine-textured soil and high impervious cover, a dry regime required 1,550 m^2 of floor per ha of drainage area. Using the SCS method for a 10-year design storm, the effect on downstream peak flood flows was tested for infiltration of either the entire volume of post-development runoff or the difference in runoff attributable to development. The results showed that either capacity was sufficient to suppress downstream peak flow during the 10-year storm to predevelopment levels, so the choice was left up to the lot developers. Infiltrating the entire volume would require a large basin, but would eliminate all costs for channels and culverts to carry design-storm runoff through and off the lot. Infiltrating the difference in volume would require channels or culverts to carry the residual runoff, but it would permit small basins.

COMPOSITE SITE EVALUATION

When the water balance effects of an infiltration basin are added to those of the drainage area it controls, the results show the composite effects of urban land use and stormwater control on disposition of water into the environment. They can be evaluated by comparing estimates for alternative land uses and stormwater controls.

Design could aim to produce no increase in direct runoff (ΔDRo) in the long-term water balance, compared to predevelopment levels. This is as valid a way to characterize downstream flooding effects as the rates, durations and volumes of isolated design storms. The long-term water balance gives a composite index of all storm events over time, as opposed to a detailed description of any specific event.

Design could aim to produce no decrease in base flow (ΔBRo). This would be a penetrating way to control effects on stream base flows, aquatic ecosystems, and ground water levels, in order to address a range of sensitive resources that have not otherwise seemed amenable to deliberate management or restoration.

Ferguson, Ellington and Gonnsen (1991) elaborated on the Minnesota method by dividing an urban site into four connected drainage zones: impervious, pervious receiving runoff from impervious areas, pervious not receiving runoff, and reservoir. A monthly water balance is calculated for each zone; the results are transferred to the next zone in the drainage se-

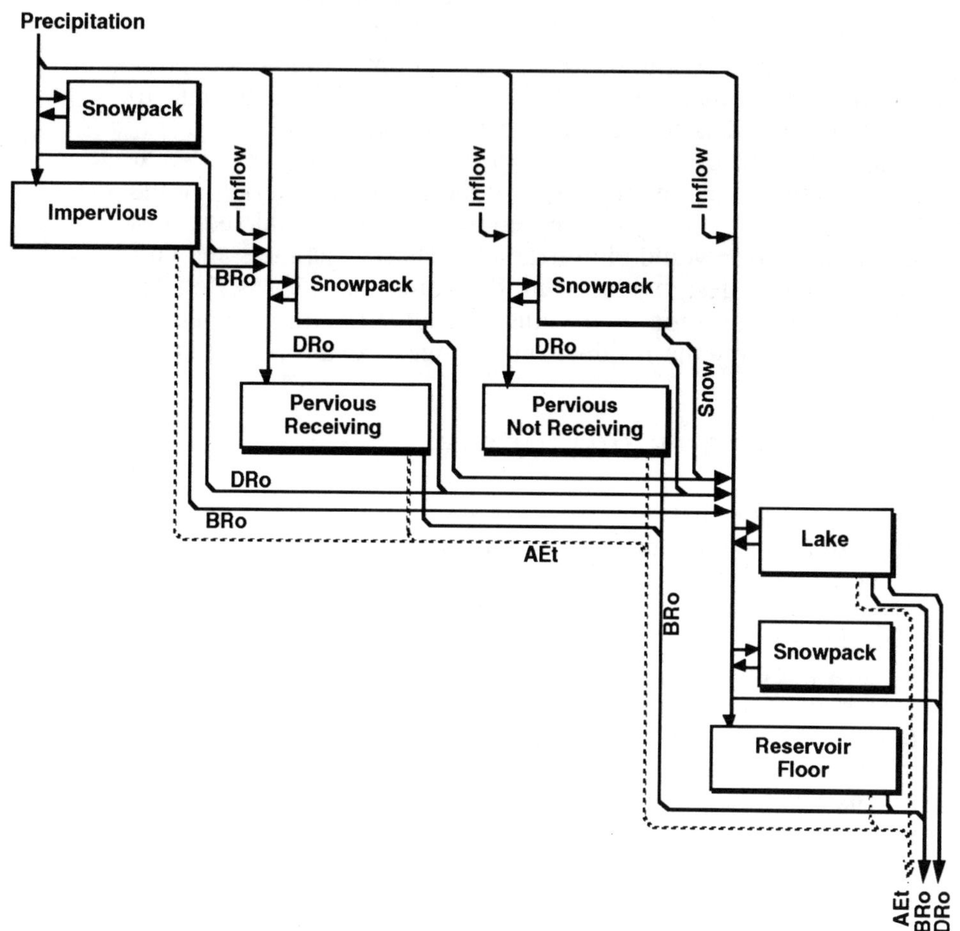

Figure 4.5 Flows and storages in Ferguson, Ellington and Gonnsen's (1991) elaboration of the Minnesota method for composite site and reservoir evaluation.

quence. Figure 4.5 shows the flows and storages in the model. In the figure, rectangles represent storages, solid lines represent flows of liquid water and dashed lines represent evapotranspiration. All flows move from the upper left of the diagram to the lower right. At the bottom of the drainage sequence is a reservoir (an infiltration basin) which is given its own water balance. The reservoir may be specified as either on the land surface, and susceptible to precipitation and evapotranspiration, or in an underground chamber, where precipitation and evapotranspiration do not occur. Lake storage (ponding) can occur above the soil surface. By setting different infiltration rates and reservoir dimensions, a user could induce a dry, ephemeral or wet regime, or the reservoir could be reduced to a nonfunctional point through which all flows pass unaltered. The total water balance of the site or drainage area is the net discharge after calculation of all the zones. Discharges of direct runoff, base flow and evapotranspiration are summed for the site as a whole.

The Philadelphia firm of Tourbier and Walmsley (J. Toby Tourbier, personal communication, October 1992) has proposed to municipalities in Pennsylvania a simple standard and procedure for new urban developments to emulate the predevelopment water balance. The predevelopment direct runoff/precipitation (DRo/P) ratio is found for the 2-year storm and then applied to annual precipitation to estimate annual direct runoff. The portion of precipitation that does not directly run off is assumed to be infiltration. An infiltration basin is added to the proposed development with an initial trial volume equal to 1/30 of the maximum monthly average precipitation. The average annual water balance of the basin is then analyzed. If annual infiltration from the basin is less than the predevelopment infiltration of the drainage area, then the basin is enlarged and the water balance recalculated. This procedure does not take into account the seasonal variation in DRo/P that occurs over the course of a year on many sites. However, taking that variation into account in the annual water balance requires a more involved calculation, which may be difficult to get every local designer to apply in daily practice.

CONTINUOUS SIMULATION ANALYSIS

Another family of hydrologic models, continuous simulation models, apply long series of meteorologic input data, the results of which can be analyzed statistically to evaluate long-term hydrologic behavior. The input data are typically on a daily time increment, sometimes with shorter time increments during precipitation and runoff events. A water balance is conducted in each time increment, with the results carried into the next time step. Many years' worth of data must be entered in order for statistical results such as recurrence intervals to be statistically valid.

Continuous simulation models are typically demanding of computing resources, but modern ones incorporate many refinements in hydrologic modeling. SWMM (Huber et al., 1975) has been widely used for urban applications; it consists of various modules or blocks which are applied to specific tasks of runoff and pollutant estimation and transport, pollutant modification in holding ponds, and the fate of pollutants in receiving surface water bodies. CREAMS (Smith and Williams, 1980) has been applied in non-urban environments; it uses daily data to estimate evaporation, direct runoff, base flow, and the transport of pollutants through surface and subsurface flows. A variation upon CREAMS is Williams, Nicks and Arnold's (1985) SWRRB, which can accommodate subareas having different soils and land uses. Another sophisticated model with specifically urban components is that of the U.S. Geological Survey (Smith and Alley, 1981), which distinguishes impervious areas that are directly connected to the channel drainage system from those that drain onto pervious ground, where the runoff supplements precipitation. In Grimmond, Oke and Steyn's (1986) distinctive model, the land surface is divided into impervious, pervious irrigated, and pervious unirrigated zones; irrigation is added to precipitation on the irrigated zone, and urban enhancement of evapotranspiration is taken into account.

Continuous simulation analysis promises greater water balance accuracy than applications of average monthly data, by taking into account temporary extremes in precipitation, runoff and evaporation not reflected in monthly averages. The size of basin needed during an extreme day might differ from that suggested by an analysis of monthly values. The most sophisticated of continuous simulation models analyze individual runoff events in as much detail as design storm models, yielding both the long-term patterns and the design storm extremes from single model runs. As computing resources and large data sets become increasingly available to practitioners, continuous simulation models are likely to become more common in design of infiltration and other kinds of stormwater management.

Driscoll et al. (1986) proposed a surrogate method for analyzing variations in long-term results without the brute computing force required for continuous simulation. It is based on application to each specific infiltration site of previously derived statistical summaries of regional precipitation data. Runoff is estimated and the basin is sized for the mean of precipitation events. The basin is then analyzed for relative performance over a wider range of events using the coefficient of variation of precipitation. The coefficient of variation may not be a valid statistic of precipitation data, because precipitation is seldom normally distributed. In addition, this surrogate method fails to take into account seasonal variations in runoff/precipitation ratios, the accumulation of low but continuous background flows, or any aspect of the water balance other than the amount of runoff entering the basins.

REFERENCES

Amoozegar, A., and A.W. Warrick, 1986, Hydraulic Conductivity of Saturated Soils: Field Methods, pages 735-770 of *Methods of Soil Analysis, Part I. Physical and Mineralogical Methods*, Agronomy Monograph no. 9, second edition, Madison: American Society of Agronomy—Soil Science Society of America.

Bouwer, Herman, 1966, Rapid Field Measurement of Air Entry Value and Hydraulic Conductivity of a Soil as Significant Parameters in Flow System Analysis, *Water Resources Research* vol. 2, no. 4, pages 729-738.

Bouwer, Herman, 1986, Intake Rate: Cylinder Infiltrometer, pages 825-844 of *Methods of Soil Analysis, Part I. Physical and Mineralogical Methods*, Agronomy Monograph no. 9, second edition, Madison: American Society of Agronomy—Soil Science Society of America.

Debo, Thomas N., (1994), *Infiltration Structure Design User's Manual*, Washington, D.C.: National Stone Association.

Driscoll, Eugene, Dominic DiToro, David Gaboury and Philip Shelley, 1986, *Methodology for Analysis of Detention Basins for Control of Urban Runoff Quality*, EPA 440/5-87-001, NTIS Document No. PB87-116562, Washington, D.C.: U.S. Environmental Protection Agency Office of Water Regulations and Standards.

Duchene, Michael, and Edward A. McBean, 1992, Discharge Characteristics of Perforated Pipe for Use in Infiltration Trenches, *Water Resources Bulletin* vol. 28, no. 3, pages 517-524.

Dunne, Thomas, and Luna B. Leopold, 1978, *Water in Environmental Planning*, San Francisco: Freeman.

Engstrand, Daniel, 1983, *Retention Ponds: Analysis and Design of Ponds without Outlets*, St. Paul: Minnesota Department of Transportation.

Ferguson, Bruce K., 1990a, Role of the Long-Term Water Balance in Design of Multiple-Purpose Stormwater Basins, pages 161-170 of *CELA '89: Proceedings of the Annual Conference of the Council of Educators in Landscape Architecture*, Sara Katherine Williams and Robert R. Grist, editors, Washington, D.C.: Landscape Architecture Foundation.

Ferguson, Bruce K., 1990b, Role of the Long-Term Water Balance in Management of Stormwater Infiltration, *Journal of Environmental Management* vol. 30, pages 221-233.

Ferguson, Bruce K., M. Morgan Ellington and P. Rexford Gonnsen, 1991, Evaluation and Control of the Long-Term Water Balance on an Urban Development Site, pages 217-220 of *Proceedings of the 1991 Georgia Water Resources Conference*, Kathryn J. Hatcher, editor, Athens: University of Georgia Institute of Natural Resources.

Grimmond, C.S.B., T.R. Oke and D.G. Steyn, 1986, Urban Water Balance. I. A Model for Daily Totals, *Water Resources Research* vol. 22, no. 10, pages 1397-1403.

Hannon, Joseph B., 1980, *Underground Disposal of Storm Water Runoff, Design Guidelines Manual*, FHWA-TS-80-218, Washington, D.C.: U.S. Federal Highway Administration.

Huber, W.C., J.P. Heaney, M.A. Medina, W.A. Peltz, H. Sheikhj and G.F. Smith, 1975, *Storm Water Management Model User's Manual, Version II*, EPA-670/2-75-017, Athens, GA: U.S. Environmental Protection Agency.

Klute, A., and C. Dirksen, 1986, Hydraulic Conductivity and Diffusivity: Laboratory Methods, pages 687-734 of *Methods of Soil Analysis, Part I. Physical and Mineralogical Methods*, Agronomy Monograph no. 9, second edition, Madison: American Society of Agronomy—Soil Science Society of America.

Maryland Water Resources Administration, 1984, *Standards and Specifications for Infiltration Practices*, Annapolis: Maryland Department of Natural Resources.

Paul Thiel Associates Limited, 1980, Subsurface Disposal of Storm Water, pages 175-193 of *Modern Sewer Design*, Washington, D.C.: American Iron and Steel Institute.

Paine, John E., 1990, *Stormwater Design Guide, Portland Cement Pervious Pavement*, Orlando: Florida Concrete and Products Association.

Prill, Robert C., and David A. Aronson, 1978, *Ponding-Test Procedure for Assessing the Infiltration Capacity of Stormwater Basins*, Nassau County, New York, Water-Supply Paper 2049, Washington, D.C.: U.S. Geological Survey.

Rawls, W.J., D.L. Brakensiek and K.E. Saxton, 1982, Estimation of Soil Water Properties, *Transactions of the American Society of Agricultural Engineers* vol. 25, no. 5, pages 1316-1320 and 1328.

Smith, Peter E., and William E. Alley, 1981, Rainfall-Runoff-Quality Model for Urban Watersheds, in *Applied Modeling in Catchment Hydrology*, Vijay P. Singh, editor, Littleton, Colorado: Water Resources Publications.

Smith, R.E., and J.R. Williams, 1980, Simulation of the Surface Water Hydrology, in *CREAMS: A Field-Scale Model for Chemicals, Runoff, and Erosion from Agricultural Management Systems*, Conservation Research Report No. 26, Walter G. Knisel, editor, Washington, D.C.: U.S. Department of Agriculture.

Stahre, Peter, and Ben Urbonas, 1989, Swedish Approach to Infiltration and Percolation Design, pages 307-322 of *Design of Urban Runoff Quality Controls*, Larry A. Roesner, Ben Urbonas and Michael B. Sonnen, editors, New York: American Society of Civil Engineers.

Thelen, Edmund, and L. Fielding Howe, 1978, *Porous Pavement*, Philadelphia: Franklin Institute Press.

U.S. Environmental Protection Agency, 1981, *Process Design Manual for Land Treatment of Municipal Wastewater*, Cincinnati: Center for Environmental Research Information.

U.S. Soil Conservation Service, 1968, *A Method for Estimating Volume and Rate of Runoff in Small Watersheds*, SCS-TP-149, Washington, D.C.: U.S. Soil Conservation Service.

Urban, James B., and William J. Gburek, 1980, Storm Water Detention and Groundwater Recharge Using Porous Asphalt—Experimental Site, pages 81-87 of *International Symposium on Urban Runoff*, Lexington: University of Kentucky.

Wanielista, Martin P., and others, 1981, *Stormwater Management Manual*, Tallahassee: Florida Department of Environmental Regulation.

Washington State Department of Ecology, 1992, *Stormwater Management Manual for the Puget Sound Basin (The Technical Manual)*, Olympia: Washington State Department of Ecology.

Way, Douglas, 1978, *Terrain Analysis, A Guide to Site Selection Using Aerial Photographic Interpretation*, second edition, Stroudsburg: Dowden, Hutchinson and Ross.

Weaver, Robert J., 1969, *Recharge Basins for Disposal of Highway Storm Drainage: Theory, Design Procedure, and Recommended Engineering Practices*, Research Report 69-2, Albany: New York State Department of Transportation, Engineering Research and Development Bureau.

Williams, J.R., A.D. Nicks and J.G. Arnold, 1985, Simulator for Water Resources in Rural Basins, *Journal of Hydraulic Engineering* (American Society of Civil Engineers) vol. 111, no. 6, pages 970-986.

5

Infiltration Performance

Performance of stormwater infiltration in a given site development is defined by the objectives of the project, as it is in any other human endeavor. Meaningful stormwater management objectives must be founded on an understanding of specific conditions both within the limits of the site and in the larger landscape systems of which on-site phenomena are part, including surrounding urban neighborhoods, upstream and downstream watersheds and underlying landforms and soils.

Stormwater infiltration is capable of advancing a broad range of water resource and environmental purposes encompassing water purity, ground and surface water reserves, flood prevention, stream channel stability, ecosystem health, and the scenic beauty of stream valleys. It is the only known method of reducing runoff volumes, restoring ground water recharge, augmenting stream flow and preserving the hydroperiod of downstream wetlands. It is surprisingly more effective than detention in controlling urban flooding and drainage problems. Infiltration surfaces and basins need not cost more to construct than any other effective stormwater management structures designed to equivalent standards. The locations, sizes, shapes and constructions of infiltration surfaces and basins influence the degree to which they accomplish these things.

Infiltration is the most complete of stormwater management approaches. There is no reason not to consider using it as a supplement to or replacement for any other method of stormwater management.

WATER QUALITY

America's success in controlling point pollution sources during the 1970s and 1980s exposed to view the importance of non-point sources such as urban stormwater. The 1986 amendments to the federal Water Quality Act mandated control of non-point pollution, making the Environmental Protection Agency an active partner in stormwater management for the first time.

Water's constituents can be dissolved or suspended (particulate) in form. The composition and quantity of constituents in urban runoff vary with precipitation quality, watershed mineralogy and all aspects of land use. Some constituents can harm humans directly when they reach water supplies or recreational lakes or are concentrated in food fish. Others, when sufficiently concentrated, can upset aquatic ecosystems.

Properly constructed infiltration surfaces and basins are traps and transformers of pollutants received with inflowing waters. They need not generate pollutants internally. This was confirmed by Gburek and Urban (1980), who monitored organic and inorganic constituents of percolate under a porous asphalt pavement in Pennsylvania. The pavement had no traffic or inflowing runoff, so water quality could be influenced only by the precipitation, the pavement material and the underlying soil. They found all concentrations in the underlying ground water well within drinking water standards; the pavement percolate did not contribute to ground water contamination. This would be expected based on the inertness of most aggregates and the near-insolubility of bituminous binders.

In Pennsylvania, Watschke and Mumma (1989) obtained analogous results for infiltration through grass. They observed experimental turf plots seeded in various mixtures of Kentucky bluegrass (*Poa pratensis*), fescue (*Festuca rubra*), perennial ryegrass (*Lolium perenne*), and annual ryegrass (*Lolium multiflorum*). Quantity of runoff was insignificant due to the soil porosity maintained by vegetation and organic matter. They found that dense, healthy turf grass infiltrated and metabolized soluble nutrients and pesticides, reducing outflows in soil percolate.

Nightingale (1987a and 1987b) confirmed the general trap effectiveness of infiltration basins that had been in operation for up to 20 years. He found that vegetated and unvegetated surface basins in Fresno, California were trapping metals and other constituents in the near-surface soils. Inflowing pollutants were not significantly contaminating moisture percolating through the alluvial soils. Concentrations of constituents in the ground water under the basins were similar to those in ground water elsewhere in the region.

Mechanisms of quality alteration

Infiltration exploits the physical, chemical and biological powers of soil to trap, attenuate and transform pollutants before they reach aquifers or streams. Detailed reviews of processes and magnitudes of soil's effects on water quality were presented by Chang and Page (1985), Gerba and Goyal (1985), McCarty, Rittmann and Reinhard (1985), and Nellor, Baird and Smyth (1985).

The following specific processes alter the quality of infiltrating water during ponding and subsurface travel: filtration, adsorption, biodegradation, growth of microorganisms, chemical oxidation and reduction, chemical precipitation and dilution, volatilization, and photochemical reactions (O'Hare et al., 1986, p. 20).

Where plants are present, they absorb nutrients. They then excrete or lose small amounts during the growing season, and release a large proportion at senescence. Harvesting plants before senescence permanently removes some of the constituents from the system. Constituents tied up in organic litter, and eventually in sediment, may represent essentially permanent storage.

Soils vary in their ability to filter and adsorb. Cation exchange capacity (CEC) and anion exchange capacity (AEC), which measure a soil's capacity to exchange ions with infiltrating water, are determined mostly by soil clay and organic matter. In coarse-textured soils, water moves rapidly through the inert soil with little opportunity for soil-water contact, filtration or biodegradation. Fine-textured and organic soils have small pores for physical filtration, electrochemically active particle surfaces, and great surface area for ion exchange and microorganism habitat, producing great soil-water contact, residence time and ion exchange (O'Hare et al., 1986, p. 41).

The degree of purification increases with distance of travel through a soil. Dissolved solids in infiltrated water decrease with subsurface travel distance (Bianchi and Muckel, 1970, p. 51). Many chemicals and nearly all pathogenic bacteria are filtered out within 1 to 3 m of vertical percolation or 15 to 60 m of lateral movement (Jackura, 1980, p. 11). For this reason it has been recommended in California that the floor elevations of infiltration basins maintain a 3 m vadose zone above the water table (Jackura, 1980, p. 13; O'Hare et al., 1986, p. 40). Soil filtration is bypassed where drainage wells inject runoff directly into open bedrock fractures and faults.

Where adsorption is a major removal mechanism, the soil's adsorption capacity may become exhausted, such that additional inflowing constituents are allowed to continue further along the flow path (O'Hare et al., 1986, p. 41). The soil remaining in the flow path would then continue to adsorb and store pollutants many years before they reached the water table or other point of concern. For complex compounds such as pesticides, the long residence time resulting from adsorption provides significant opportunity for biodegradation.

In certain arid areas, infiltrating water may acquire soluble salts from the soil (Bianchi and Muckel, 1970, p. 51). The salinity of the water decreases after enough water has passed through the profile to bring the soil into chemical equilibrium with the applied water.

Many of infiltration's pollutant removal mechanisms may be ineffective where soils are extremely permeable and contain little clay or organic matter. In these conditions water percolates directly to the water table without filtering, adsorption or residence time in the vadose zone. For this reason the Washington

State Department of Ecology (1992, chapter III-3) requires that the soil in the first 45 cm under a basin floor have hydraulic conductivity not exceeding 6 cm/h and cation exchange capacity of at least 5 milliequivalents per 100 g of dry soil. Where the native soil does not meet these criteria, an "aquitard" of finer-textured soil may be added as a liner to a basin's floor. Alternatively, runoff may be pretreated in a settling basin before entering an infiltration basin, or inflow to an infiltration basin can be selected from roofs only.

Some aspects of pollutant removal may be bypassed where a wet basin regime prevents aerobic microbial decomposition or where there is insufficient depth of vadose zone between the basin floor and the water table. For this reason the Washington State Department of Ecology (1992) requires that the 6 month, 24-hour storm be infiltrated within 24 hours and that the basin floor be located at least one meter above the water table, bedrock or any impermeable layer. Analyzing a basin with the long-term water balance would be another way to assure a year-round dry regime. The Maryland Water Resources Administration (1984) requires that basin floors be at least 1.2 m above the water table. In California (Jackura, 1980) it has been suggested that the bottoms of drainage wells be at least 3 m above the water table.

Treatment of first flush

In some regions, rainstorms tend to be separated by a few days or more. For example, in the mid-Atlantic area of the United States, the average duration between storms is 3.3 days (Maryland Water Resources Administration, 1986). Oils, solids and other pollutants accumulate on pavements between storms. These constituents are flushed away by the first runoff that occurs during each storm, no matter how big the total storm after the first flush has occurred. During storms of long duration, the first flush of runoff has a high concentration compared to runoff later in the storm; the quality of runoff gradually approaches that of the precipitation (Ku and Simmons, 1986). A large proportion of the constituent load is carried in a relatively small proportion of total runoff volume.

Due to the first flush effect, significant pollutant removal is possible with infiltration designed only for a small quantity of runoff, for instance the first 1.25 cm of runoff or all storms up to the 2-year storm. As long as availability of basin capacity to capture the storm flow is confirmed using the long-term water balance, the first flush of large storms and the entire runoff volume of small storms can be infiltrated and treated in the soil.

Urban pollutants other than sediment originate largely on the impervious portions of watersheds. The Maryland Water Resources Administration (1986) calculated that, in its area, 1.25 cm of runoff captured from the impervious areas alone would lead to high levels of removal. Sixty to 65 percent of the total annual runoff from the impervious areas would be captured, including all of the runoff from 70 to 75 percent of storm events each year, and containing 85 percent of the annual pollutant load.

The volume of storm flow in excess of that captured by an infiltration basin continues on down the surface flow route. In an "on-line" basin, inflow enters at one end of the basin and overflow exits from the other end. In a large storm, this type of construction may reduce the basin's effective capture of the first flush: water arriving late in the storm period could displace the initially captured first flush, pushing the first flush out the overflow faster than it can infiltrate. An alternative is an "off-line" arrangement, in which low flows are diverted by a drop inlet or low weir out of a channel into a nearby infiltration basin, while high flows continue down the channel. Only the first flush enters the basin. Without displacement by later flow, the first flush has a full opportunity to infiltrate.

Accumulation of constituents in infiltration basins

Infiltration basins tend to concentrate the contaminants from their drainage areas onto small soil areas and to trap them there. Accumulation of constituents on basin floors confirms that the basins are effective pollutant filters.

Detailed studies of lead and other metals have illustrated the accumulation of constituents in specific basins. Leaded gasoline and wearing automobile parts have been important sources of lead in urban runoff. Lead has accumulated in soil and vegetation near heavily traveled streets, in decreasing amounts with distance from the roadway. Nightingale (1978) sampled

16 infiltration basins in Fresno, of which 7 were planted with grass and 5 were not; 3 were developed for recreation. The metals arriving at the basins in runoff were mostly in the suspended form. The results showed that a large amount of lead accumulated in the first 5 cm of soil, mostly in association with the soil's clay and organic matter portions. In the recreational (baseball diamond) area of one basin, lead in the surface (0-15 cm) layer had increased about 16 times, or 30 to 35 mg Pb/kg soil/year, over background levels in the soil. The amount of lead decreased in the 5 to 15 cm depth interval and reached natural background levels in the 15 to 30 cm interval. Where sand filters were used at recharge wells within the basins, the sand trapped clay carrying lead and prevented it from entering the wells. Zinc and possibly copper were also found to be elevated in the surface layer compared with background levels.

Accumulated metals in the surface soil could eventually limit multiple use or require mechanical removal of the contaminated surface soil from the basin floor, although no stormwater infiltration basin has yet been reported to reach toxic pollutant levels. The concentrations of metals observed in Fresno (Nightingale, 1978) after 2 to 12 years of basin operation did not constitute a hazard to the use of the basins for further recreation or ground water recharge.

Accumulation of organic chemicals such as pesticides has also been studied in detail. Nightingale (1987c) measured a wide range of organics in the soils of five basins in Fresno. Although pesticides were known to be present in Fresno's runoff, only chlordane was found consistently in basin soils. It reached the moderate level of 2.7 mg/kg soil and decreased with depth to less than 0.03 mg/kg below 24 cm. Pollution of the soil by other organic compounds was insignificant. There was no evidence that man-made organic compounds had an unfavorable impact on the plants in the basins or on the use of turfed basins for recreation during dry periods.

Exchanging another effective form of runoff quality control, such as artificial wetlands, for infiltration would not avoid accumulation; it would only change the mechanism of accumulation. Accumulation somewhere in the environment can be avoided only by reducing the generation of pollutants at their sources in automobiles, construction sites, lawn fertilizers, etc. The decline in use of lead in gasoline since about 1970 has in fact reduced lead pollution at one source.

Solids

Suspended sediment is probably the most common urban runoff pollutant. Soil particles originate on construction sites, and in drainage swales and stream channels eroding in response to stressful pulses of urban stormwater. Other suspended particles include glass, asphalt, stone, rubber and rust from cars, and decomposing building materials and pavements. Suspended solids can reduce light penetration into streams and lakes, create slicks on urban walkways and pavements, clog storm sewers, aggrade streams, fill ponds and reservoirs, and carry chemical constituents.

During infiltration of water into the soil, almost all solids are removed by filtration and adsorption in the initial centimeters of soil. Ground water tends to be nearly free of suspended solids because of the filtering and adsorptive properties of soil (Bianchi and Muckel, 1970, p. 51).

Microorganisms

All surface waters and surface soils contain large, active populations of bacteria, fungi and other organisms that exist on supplies of organic material (Bianchi and Muckel, 1970, p. 52-53). They are particularly active in the warm season.

Pathogenic microorganisms can include bacteria, viruses, protozoa and parasitic worms (O'Hare et al., 1986, p. 42). Coliform bacteria are commonly used as an indicator of potential sewage pollution and associated pathogens. In runoff, most coliform bacteria are native soil organisms that are washed off soil particles by moving water (Ku and Simmons, 1986). Fecal coliforms and fecal streptococci, however, come from warm-blooded animals. In urban areas, fecal bacteria are commonly derived from dogs, cats, rodents, birds and other small animals. They may be concentrated in the runoff from restaurants, dumpsters and service areas where trash is handled. In some watersheds, sanitary sewer leaks and overflows are critical sources of coliforms.

The removal of microorganisms from water during infiltration is determined by their extinction in ponded water and their retention and extinction in the soil matrix (O'Hare et al., 1986, p. 42).

Ponding and soil saturation during infiltration vary the availability of oxygen and nitrogen for microbiotic metabolism (Bianchi and Muckel, 1970, p. 52-53). Ponding shuts off the soil environment from atmospheric oxygen and shifts the soil population toward organisms that can exploit other sources of energy.

The size of microorganisms is such that they are effectively filtered as water moves through most soils, usually within a few meters of travel (Bianchi and Muckel, 1970, p. 52-53; Jackura, 1980, p. 11; O'Hare et al., 1986, p. 20). In the case of viruses, the main method of attachment is by adsorption. The removal of microorganisms, including viruses, during passage of water through most soils is believed to be virtually complete. Despite the highly permeable soils typical of most of Long Island, Ku and Simmons (1986) found the Island's infiltration basins effective in removing bacteria from stormwater before it reached the water table. To assure removal of bacteria before runoff reaches the water table, the Maryland Water Resources Administration (1984) specifies at least 1.2 m clearance from the floor of a basin to the seasonal high water table.

Microorganisms filtered in the soil can be as effective in plugging soil pores as clay and silt particles. An improvement in water quality may come at the expense of infiltration rate, as it may for other solids. However, unlike other solids, microorganisms can be removed from soil pores by oxidation (Bianchi and Muckel, 1970, p. 52-53).

Metals

Metals in runoff can originate from cars and trucks. Lead can be high at highways and commercial parking lots. Although the phasing out of leaded gasoline has reduced one source of lead, lead continues to be released from other sources such as automobile wheel weights and the wear of certain kinds of tires. Other metals originate from decomposition of pavement and construction materials, including oxidized iron, zinc and copper from roofs, culverts, nails and painted surfaces. Metals can be leached out of some fertilizers.

At five infiltration basins on Long Island, Ku and Simmons (1986) compared concentrations of constituents in precipitation, runoff entering the basins, soils in basin floors, soil moisture below basin floors, and ground water beneath basins and in the surroundings. Where they found large inflows of chromium or lead, large degrees of removal also occurred before the water reached the water table, sometimes leaving concentrations in ground water of less than 5 percent of those in the basin's inflow. Metals in particulate form were filtered out by the soil. Metals in dissolved form were adsorbed onto near-surface soil particles.

Wigington et al. (1983) confirmed that heavy metals such as lead accumulate in the top few centimeters of soil and that movement downward through the soil is limited. The quantities and rates of accumulation in basin soils are highly variable and dependent on the land use, the microtopography, the soil type, and the residence time of runoff in the basin.

Nutrients

Nutrients, particularly nitrogen and phosphorus, originate in fertilizers concentrated at intensely maintained golf courses, residences and office parks. They are leached from dumpsters and service areas where trash is handled, and are carried in automobile-related organic compounds concentrated at highways and commercial parking lots.

Excessive nutrient concentrations lead to high biological and chemical oxygen demands (BOD and COD), which measure the composite amount of organic substances able to be oxidized through aerobic biochemical processes. High BOD and COD indicate oxygen starvation and lead to ecosystemic imbalance.

Nitrogen is versatile and mobile in the environment, taking on different compound forms and physical states with different histories of aerobic and anaerobic biochemical processes. Excessive nitrogen is a direct human health hazard.

In infiltration basins, with ponding and depth in the soil, aerobic nitrogen-fixing organisms give way to anaerobic denitrifying organisms. Nitrate is reduced to nitrogen gas which is subsequently lost to the atmosphere. Removals of 50 to 80 percent of nitrogen from runoff water have been observed, varying with the flooding and drying regimes of the basins (O'Hare et al., 1986, p. 40).

In Sweden, wastewater infiltration basins have been managed as intermittently submerged wetlands, planted with sweetgrass (*Glyceria maxima*). They are referred to as infiltration wetlands (Wittgren and Sundblad, 1990). Large losses of nitrogen to the air have been observed during residence in the basins. Both aerobic and anaerobic soil processes are able to transform nitrogen during submergence cycles, due partly to algal photosynthesis in ponded water.

Under basins on Long Island, after some storms, Ku and Simmons (1986) found nitrogen levels in the ground water greater than those in the stormwater entering infiltration basins. The nitrogen had apparently entered the regional ground water from other sources such as septic tanks, cesspools and lawn fertilizers. The infiltration basins were apparently diluting nitrogen in the ground water with fresh water, where the aquifer was already polluted with nitrogen from other sources.

Most phosphorus in urban runoff is transported on the surface of suspended soil particles. Infiltration basins filter and adsorb these solids out of runoff. Most soils have large capacities to adsorb phosphorus. Dissolved phosphorus is precipitated in soils as calcium phosphate or magnesium ammonium phosphate. The degree of phosphorus removal depends partly on the travel distance through the soil: distances of 9 m have produced 50 percent reductions, and distances of over 100 m have led to 90 percent reductions (O'Hare et al., 1986, p. 40). The degree of removal can be lower in basins that are ponded for a long time: in anaerobic conditions, phosphorus can be remobilized and continue to move along the water's flow path.

Organic compounds

Organic compounds include a huge array of complex chemicals, mostly derived from petroleum. They are organic in the chemical sense, being structured around hydrocarbons. Major sources are oil and gasoline, concentrated at highways, commercial parking lots and any place where used crankcase oil drains into surface channels. Other sources are herbicides and pesticides applied to golf courses and other intensely maintained landscapes; these chemicals represent a direct human health hazard. Most pesticides are nearly insoluble in water (Ku and Simmons, 1986); in flowing runoff they can be mixed with the water or sorbed on sediment particles.

Leonard et al. (1976) reviewed the processes acting on pesticides in soil and water environments. In infiltration basins, dissolved organic constituents are adsorbed and biodegraded (O'Hare et al., 1986, p. 20). Hydrophobic solutes such as DDT and PCB compounds tend to be expelled from aqueous solution into strong and long-term sorption on soil surfaces. Adsorption retards solute movement and attenuates concentration fluctuations. For the few pesticide compounds that were detected in Long Island basins by Ku and Simmons (1986), relatively elevated but not toxic levels were found in basin soils, indicating their removal from water during infiltration.

Biodegradation offers the potential of permanent conversion of organic substances into harmless products (O'Hare et al., 1986, p. 20). In subsurface environments, biodegradation is accomplished mostly by microorganisms attached to solid surfaces. The rate and degree of biodegradation vary with the types of organics in the water and soil biophysical conditions such as temperature and aeration. The time required for complete biodegradation ranges from days to many years.

FLOODING

Flooding and local drainage problems are among the oldest and most common reasons for implementing any kind of urban stormwater management.

Mechanisms of flooding and drainage problems

One type of flooding and drainage problem in urban watersheds occurs in unobstructed open channels, where stream stage (water level) is a direct function of instantaneous flow rate. Urbanization typically increases peak storm flow rates. Increased peak flow rates lead to higher stream stage and increased overbank flows onto surrounding properties, hence the flooding and drainage problem. The increased peak rates are partly a result of increased storm flow volumes from urban impervious surfaces. Increased volume could cause an increased peak rate by itself, even if timing of flow is not changed. However, urban development also alters the timing of flow by increasing velocity across smooth impervious surfaces on the one hand, and attenuation in detention reservoirs and at drainage obstructions on the other. Watershed geography also affects timing: hydrographs from subwatersheds merge where tributaries join; if even a part of a tributary hydrograph overlaps the peak of the main stream, peak composite flow is increased and flooding and drainage problems worsen.

A second type of urban drainage problem occurs at flow obstructions (Ferguson and Deak, in press). Urban streams are typically obstructed at numerous culvert entrances, street crossings, fills and structures. The area upstream of an obstruction acts as a reservoir, holding arriving runoff while earlier arriving waters are still passing through the obstruction. Stage rises while water accumulates in the reservoir. Peak stage in a reservoir can be higher that that caused by peak flow rate alone in an unobstructed channel. When flow volume increases with urbanization, stage and storage rise with the prolonged inflow. Either increased volume or increased peak rate can increase stage alone, or the two can interact to produce an increase jointly. Rising of stage leads to overflowing, hence the flooding problem.

Thus increased flow volume is one of the basic mechanisms by which urbanization aggravates downstream flooding. It is a basic reason that urban development increases peak flow rates, that the hydrographs from conflowing tributaries pile up to further increased peak rates, and that stage and storage at obstructions increase to create flooding problems.

Detention, the temporary delay of surface flows in reservoirs, has no significant effect on storm flow volume. Following storage in a detention reservoir, all the runoff eventually exits through the reservoir's outlet and continues downstream. The effectiveness of detention in reducing downstream flooding and drainage problems depends on each detention basin's size and location in the watershed, and the types of flooding problems that exist at specific points in the stream system. In certain watersheds, detention at given points can actually increase peak downstream flow by extending the time over which tributary hydrographs merge (McCuen, 1974). Where urbanization has increased total volume of flow throughout a stream system, vigorous watershed-wide analysis is necessary to reduce flooding and drainage problems using detention alone.

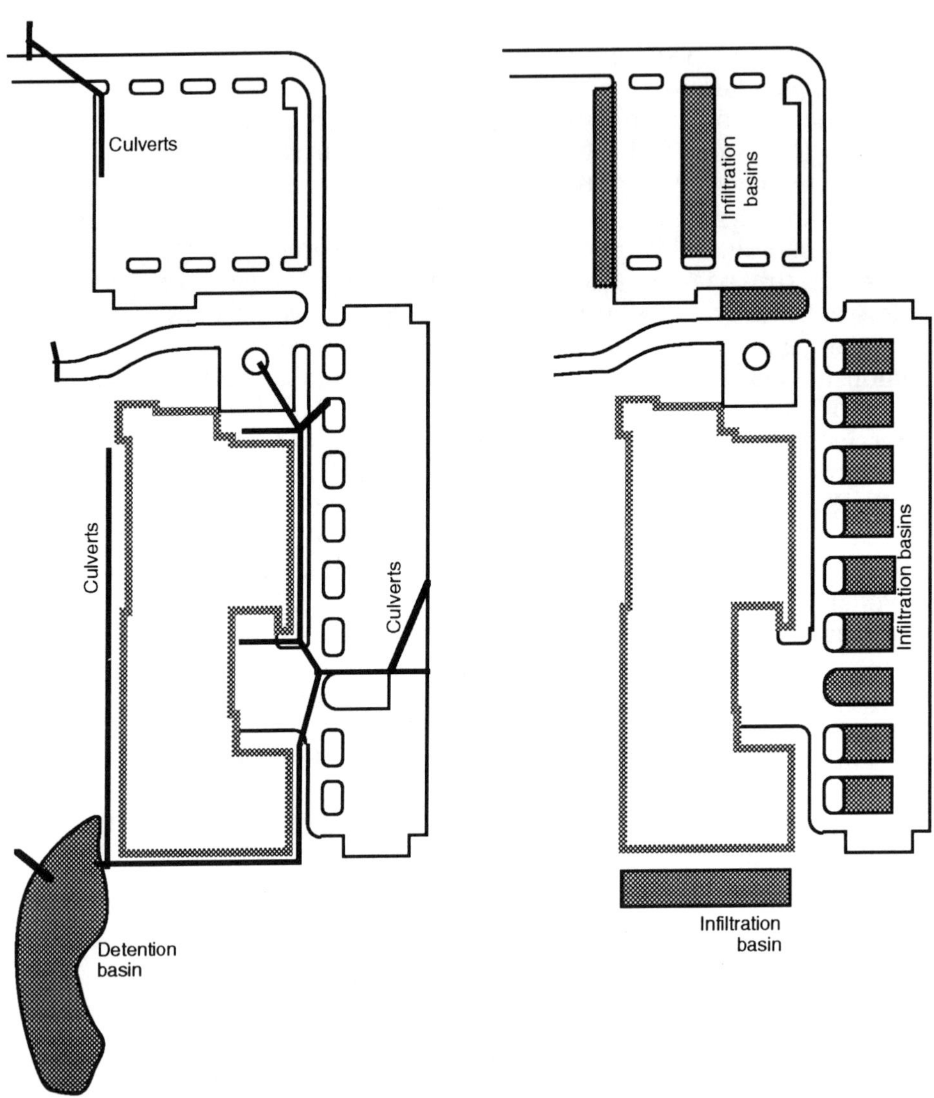

Figure 5.1 Alternative drainage plans for the industrial site modeled by Ellington (1991; Ellington and Ferguson, 1991). On the left is the detention system, with culverts leading to a detention basin. On the right is the alternative infiltration system, with subsurface infiltration basins.

Table 5.1 Hydrologic performance of alternative drainage systems on two urban sites in the Atlanta area (Ellington, 1991; Ellington and Ferguson, 1991). Flow volumes and peak rates were estimated by the SCS method.

Condition	Sum of peak discharges (m^3/s)	Velocity at peak rate of discharge (m/s)	Storm volume released (m^3)	Storm volume infiltrated through basins (m^3)
	Residential site, 50-year storm			
Undeveloped	4.7	3.8	5,040	0
Developed with detention	4.6	3.9	5,410	0
Developed with infiltration	4.6	3.8	4,490	910
	Industrial site, 100-year storm			
Undeveloped	2.3	2.2	2,540	0
Developed with detention	2.1	1.6	3,360	0
Developed with infiltration	2.2	1.7	1,930	1,530

Infiltration effects

Infiltration reduces surface flow volume during storm events. It eliminates all flow or the first portion of a flow near the source, thereby reducing the potential for flooding at all points downstream.

Infiltration does not necessarily detain any remaining flow that overflows the basin and discharges downstream. However, Figure 3.6 shows that, where more than 0.375 of the flow volume from a simple drainage area is infiltrated, peak rate of flow at the discharge (overflow) point is reduced. Thus the effect of an adequately sized infiltration basin on peak rate of flow at the discharge point can be equal to that of a detention basin. The greater the proportion of flow volume infiltrated, the greater the reduction in discharge rate.

Patton (1986) and Ellington (1991; Ellington and Ferguson, 1991) compared the flood control effects of detention and infiltration by modeling storm discharges from two urban sites near Atlanta, one residential and one industrial. Figure 5.1 shows the industrial site. Both sites had already been developed with conveyance and detention systems designed to meet existing local standards. On paper, the researchers removed the conveyance and detention systems—leaving buildings, pavements and other aspects of site layout in place—and replaced them with hypothetical infiltration systems. The infiltration basins were located to capture runoff at the source in major impervious areas, and sized as needed to meet local standards for peak rate of discharge (Ellington, 1991, p. 49 and 92-94).

Table 5.1 compares the resulting discharges predicted by the SCS method supplemented by channel and reservoir routing. On the residential site, infiltration basins sized to hold the entire volume of the 50-year storm reduced the storm flow volume to the predevelopment level and effectively controlled the peak rate. On the industrial site, located in a different county with different drainage standards, basins sized to hold the volume of either the 25- or 100-year storm reduced flow volume to the predevelopment level and effectively controlled the peak rate during the 100-year storm. Detention designed to reduce flow rate did not reduce flow volume. During the 2-year storm, which in the long run may be as damaging as large, infrequent storms, infiltration reduced the peak discharge to equal or below the predevelopment condition, while the detention systems, being designed for larger storms, were ineffective. Thus, at the point of discharge from an individual site, properly implemented infiltration is capable of suppressing flow volume and rate to chosen levels.

Ferguson, Gonnsen and Rauton (unpublished data) investigated whether Patton and Ellington's conclusion could be extended downstream amidst the converging hydrographs of tributaries when infiltration is practiced on numerous development sites in a complex watershed. They simulated the 10 year storm in a 56.5 ha watershed occupied by the Coggins Park industrial park in Athens, Georgia, using the SCS method with channel routing. They located alternative detention and infiltration basins to control runoff from each of 16 subwatersheds. They assumed that each basin would capture runoff from 90 percent of its subwatershed. In one trial, the basins were sized to capture the entire flow volume. In another, basin capacity was equal to the increase in flow volume attributable to urban development, allowing the basins to partly overflow during the design storm.

Table 5.2 shows their results at the bottom of the watershed where all flows converged. Partial infiltration adequately controlled peak flows within predevelopment levels. Full in-

Table 5.2 Volumes and rates of flow during the 10-year storm at the Coggins Park industrial site in Athens, Georgia (Ferguson, Gonnsen and Rauton, unpublished data).

Watershed condition	Runoff generated on site (m^3)	Runoff infiltrated through basins (m^3)	Runoff discharged (m^3)	Peak rate of flow (m^3/s)
Wooded (before development)	23,600	0	23,600	5.3
Developed with detention	65,900	0	65,900	8.7
Developed with infiltration of full 10 year flow volume (ΣQ_{vol})	65,900	55,800	10,100	3.1
Developed with infiltration of partial 10 year flow volume (ΔQ_{vol})	65,900	29,500	36,400	5.2

filtration overcontrolled peak flow, making it 40 percent lower than the predevelopment level. This is because, as development with infiltration was added, the subwatershed areas contributing direct runoff were reduced, while developed areas (with infiltration basins) that replaced them contributed no volume to downstream flows.

The researchers also modeled larger and smaller storms for comparison. For the larger 25-, 50- and 100-year storms both levels of infiltration were adequate when compared with predevelopment levels, although partial infiltration was not as effective as full infiltration. This was because basins designed for the 10-year storm captured only the initial portion of runoff during larger storms; some flow volume spilled over the basins and contributed to downstream flows. Basins sized only for partial infiltration overflowed sooner than the full-infiltration basins, allowing greater contributions of volume and rate to downstream peaks.

During small storms (2-, 5- and 10-year frequencies), the downstream peak flows with full and partial infiltration were identical. This was due to the occurrence of peak rate of discharge during the early part of a storm event, while an infiltration basin still has unused capacity and is capturing runoff. By the time a partial-infiltration basin is full and begins to overflow, only the residual trailing discharge is arriving at the basin from the drainage area. The basin overflows only at the low rate at which inflow is delivered to it, making very little contribution to downstream peaks.

Another way of looking at flood effects is the annual total of direct runoff. Direct runoff is generated largely by surface and near-surface processes during storm and snowmelt events. Its annual total is a composite of all runoff events year-round, as opposed to the specifically defined conditions of an isolated design storm.

Ferguson, Ellington and Gonnsen (1991) simulated the long-term water balance of the 35-ha Oconee County High School site in the Piedmont region of Georgia. The model they used was depicted in Figure 4.5. The development proposed impervious surfaces totaling 21 percent of the previously well-vegetated site area. A surface discharge plan expelled all stormwater at the surface via culverts and a detention basin. An alternative infiltration plan captured runoff from most of the impervious surfaces in stone-filled basins under parking pavements. Both plans were designed to reduce the peak rate of flow from the site's major watershed during the 25-year storm to the predevelopment level. In addition the infiltration plan was designed to reduce the total volume of flow during the same storm to its predevelopment level.

Figure 5.2 shows the average annual disposition of precipitation in each plan. Before development, with the site in pasture and woodland, evapotranspiration (Et) was about 65 percent of precipitation; the remaining 35 percent became runoff. Direct runoff was 12 percent of total runoff, or 4 percent of all water disposition. After development with the surface discharge plan, total annual runoff was greater. About half of the runoff was in direct runoff due to the increased impervious surfaces; this was about six times greater than direct runoff before development. With infiltration, most direct runoff was eliminated by capture in infiltration basins. A small increase in direct runoff over the predevelopment level remained due to some impervious surfaces, such as those around playing fields, not draining into infiltration basins.

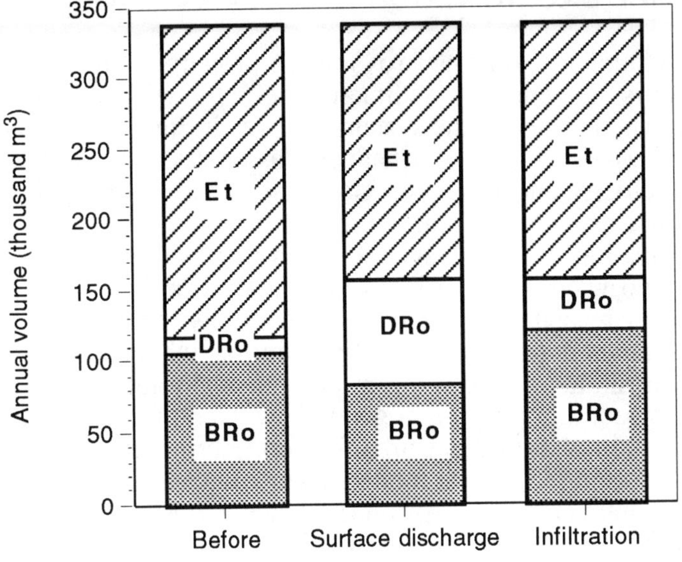

Figure 5.2 Average annual disposition of water from a 35-ha school site in Georgia, before development and under two alternative stormwater management plans. Et is evapotranspiration; DRo is direct runoff; BRo is base flow.

According to these studies, infiltration is capable of reducing volume and peak rate of storm flow at the point of discharge and consistently downstream, eliminating all causes of aggravated urban flooding and drainage problems. Infiltrating the entire postdevelopment flow volume during a design storm may not be necessary to reduce peak flow to the predevelopment rate; infiltrating only the increase in volume attributable to development may be sufficient. This partial capacity is effective at all smaller recurrence intervals and has some effect at greater recurrence intervals. Infiltration eliminates volume of runoff wherever it is practiced in the watershed; therefore it can only reduce peak flows at all downstream points. Thus infiltration, in contrast to detention, is a complete and reliable solution to overflows both in open channels and at drainage obstructions.

CHANNEL EROSION

Stream channel erosion destroys urban properties and riparian ecosystems. It generates sediment internally within a stream system, whether or not the system receives inflows of sediment from eroding upland sites. Channel erosion rate increases with stream velocity and thus with flow rate. Total volume of erosion increases with duration of eroding flows.

Urbanization in a watershed tends to aggravate channel erosion in two ways. First, it leads to increased peak storm flows and thus increased peak velocities. Second, by increasing storm flow volume, it lengthens the time during which velocity is over a threshold erosive velocity, and thus the total erosion and sediment generation during a given flow event. Given the high volume of flow following urbanization, stormwater detention tends to extend the time at moderate flow rates even further, increasing total amount of erosion (McCuen and Moglen, 1988).

Ellington's (1991; Ellington and Ferguson, 1991) modeling of sites in Atlanta compared the erosion effects of different stormwater management approaches by using velocity at the principal discharge point as an index of erosion potential.

Table 5.1 lists velocity at the peak flow rate during the design storm for each site and watershed condition. At the residential site, infiltration released a peak velocity lower than detention's and equal to that before development. At the industrial site infiltration resulted in a peak velocity slightly higher than that of detention, but still less than the undeveloped condition. Figure 5.3 shows the velocity during the 100-year storm at the industrial site. The time of infiltration discharge over the erosive threshold velocity of 1.2 m/s was only a fraction of that either before development or following development with detention. A similar conclusion was reached for small, frequent (two-year) events to which channel morphology may be more closely adapted than to the 100-year storm.

The results of Ferguson, Gonnsen and Rauton's (unpublished) study extended Ellington's conclusions downstream. On-lot infiltration reduced both volume and rate consistently in a complex watershed, allowing flow rate to return quickly to low levels without aggravating stream channel erosion.

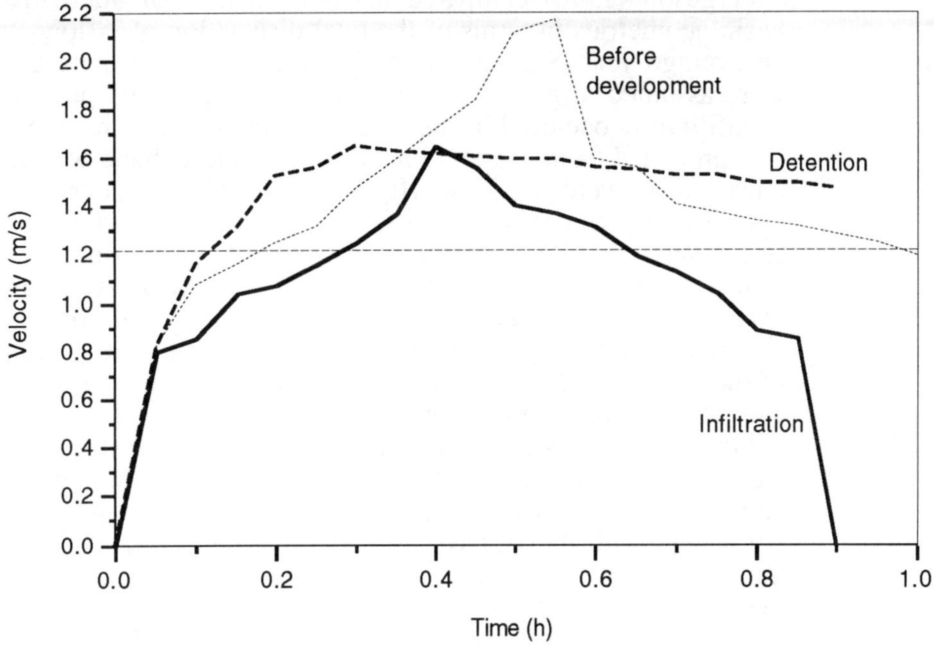

Figure 5.3 Velocity of runoff discharging during the 100-year storm from an industrial site near Atlanta, before development and under two alternative stormwater management plans (data from Ellington, 1991, p. 116).

SUBSURFACE RECHARGE

Infiltration is the only method of stormwater management capable of maintaining or restoring soil moisture and ground water reservoirs, and consequently supporting downstream base flow and wetland hydroperiod. In long-term water balance modeling, the amount of water that passes through subsurface processes is indicated by base flow. Base flow is water that has infiltrated, passed through subsurface storage and transmission processes, and later reemerged at the surface.

Ferguson (1990) compared the performance of alternative basin geometries in terms of the total disposition of inflows in an average year. Soils and rainfall typical of the Atlanta area were assumed. Figure 5.4 shows the results. High proportions of infiltration occurred in subsurface basins where stored water was protected from evaporation, as long as the volume was adequate to prevent overflow. High infiltration also occurred in open basins with volumes sufficient to prevent overflow but floors small enough to prevent great evapotranspiration; such basins tend to have wet or ephemeral regimes. Large evaporative losses occurred in open basins with large floors designed to remain dry. Large overflow losses—up to 40 percent of the annual inflows—occurred in basins with small capacities such as 1.25 cm runoff volume.

Figure 5.2 shows different amounts of base flow that resulted from different management plans in Ferguson, Ellington and Gonnsen's (1991) modeling of the Oconee school site. After development with the surface discharge plan, evapotranspiration and base flow were reduced by the decline in vegetated surfaces, despite the enhancement of soil moisture in some of the vegetated areas by runoff transfers. With stormwater infiltration, annual evapotranspiration and total runoff were the same as with surface discharge, because soil covers were the same. However, base flow was restored by the diversion of a large proportion of direct runoff from impervious areas into infiltration basins. Over the course of a year, the subsurface basins would cause more water to infiltrate after development than before, by protecting it from evapotranspiration. Thus stormwater infiltration transforms direct runoff into subsurface recharge and base flow.

These modeling results were confirmed in the field by Gburek and Urban (1980), who monitored infiltration through a subsurface basin in the form of a porous asphalt pavement near Philadelphia. The basin's capacity was sufficient for the precipitation volume of a 25-year storm. No surface runoff was produced at any time, including times when runoff was produced by an adjacent grass plot. No percolate appeared in the soil following storms of less than about 0.5 to 0.8 cm; following larger storms the percolate was from 50 to more than 100 percent of rainfall. The difference between rainfall and percolate amounts was attributed to water remaining in the unsaturated pavement profile after drainage and lost to evapora-

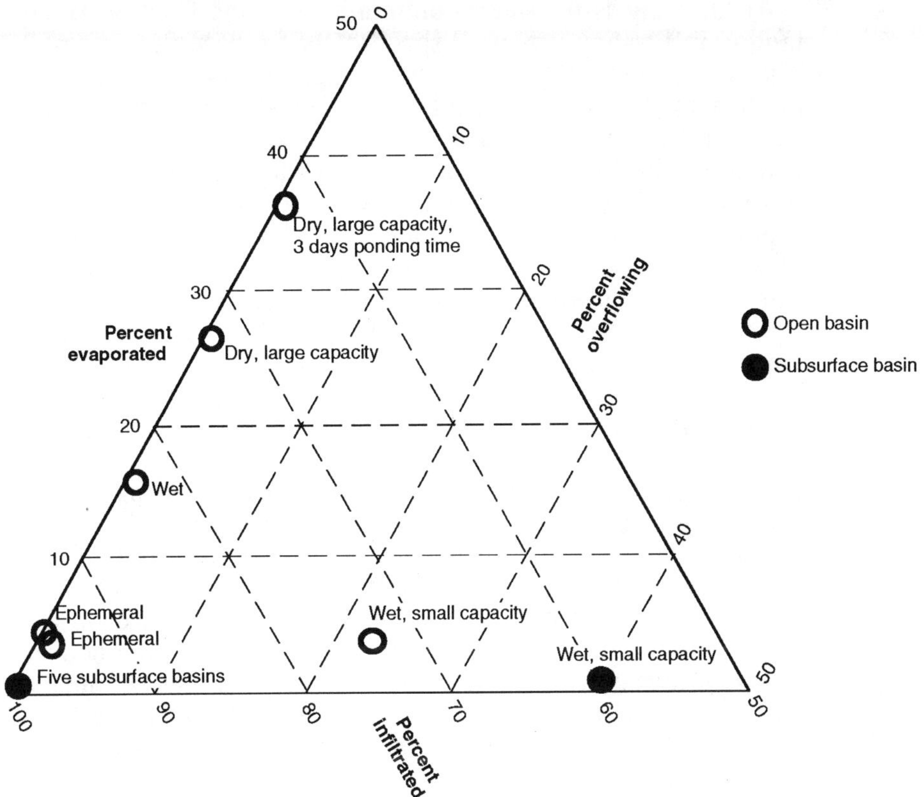

Figure 5.4 Average annual dispositions of water from infiltration basins of different volumes, floor areas, and constructions, and thus of different hydrologic regimes, at a hypothetical drainage area in the Atlanta area (data from Ferguson, 1990).

tion before the next storm began. On a monthly basis, 70 to 90 percent of the rainfall appeared as percolate below the plot in months with normal or high rainfall, and from 55 to slightly over 100 percent during months of low rainfall. Precipitation was infiltrated and recharged throughout the year, whereas that on the adjacent grass cover was recharged little or not at all during the growing season, and that on a nearby impervious pavement was recharged not at all.

At the Ring Hill Estates community in King County, Washington, infiltration was used specifically for base flow support. The development site of 130 ha contained 100 residential lots in clusters. The site straddled a ridge with steep slopes. At the bases of the slopes were wetlands sensitive to urban discharges. Under King County's stringent stormwater requirements, infiltration within the residential lots recharged ground water to support base flow into the wetlands. Also in western Washington, a ball field at the Kitsap County School doubled as an infiltration area to satisfy the required recharge in a designated aquifer recharge zone (Ed McCarthy, Geodimensions, Kirkland, Washington, personal communication).

FEASIBILITY

Feasibility of implementing stormwater infiltration in a given site can be evaluated in terms of the physical possibility of infiltration into the soil, and the cost of infiltration including land allocation, construction and maintenance.

Maryland's statewide infiltration standards of a decade ago (Maryland Water Resources Administration, 1984) defined infiltration feasibility in terms of soil texture and thus conductivity. Where soil texture was silt loam or coarser, and thus K greater than or equal to 0.68 cm/h, infiltration was defined as feasible, and developers were required to consider it in their stormwater management. Where soil was finer and less conductive, infiltration was defined as infeasible and was not considered in site development. Following Maryland's precedent, similar definitions have been adopted elsewhere and cited in secondary works (Schueler, 1987; Schueler, Kumble and Heraty, 1992), the threshold conductivity varying from one author or jurisdiction to another. This is at best an extremely superficial definition of feasibility. It does not test either physical possibility or cost for specific sites.

To say that it is not feasible to use a fine-textured soil for infiltration because it is slowly permeable is like saying that it is not feasible to use corrugated metal pipe for carrying water because corrugated metal has a high roughness factor and therefore carries little water. Corrugated metal pipes are in fact used by the thousands in every urban community, and no one claims them to be unsuited for this purpose on the grounds of their high roughness factor. Roughness factor is taken into account in pipe design: large cross-sectional area of a pipe compensates for high roughness. Using standard hydraulic formulas, the designer derives exactly the cross-sectional area needed for a given roughness and inflowing runoff. Similarly, in infiltration one cannot say that low conductivity by itself makes a soil unsuited for infiltration. Conductivity is taken into account in infiltration design: large infiltration surface area compensates for low conductivity. Using storm ponding time and water balance computation, the designer derives exactly the surface area required for a given conductivity and inflowing runoff.

There are only a few kinds of site conditions that make infiltration unworthy of even superficial consideration. These include:

• soil as impermeable as the roofs and pavements that will be placed upon it,

• toxic waste in industrial areas and saline deposits in arid areas that would leach if loaded with additional throughflow,

• steep unstable slopes, and

• proximity to basements, sensitive structural foundations, water supply wells or septic tanks.

Feasibility studies

In the mid-1980s, in the Piedmont region of Georgia (the Atlanta-Athens area), infiltration had not been attempted in the field. Its potential feasibility was questioned because of the region's high rainfall and slowly permeable soils. The soils in this area are characteristically fine-textured, well below Maryland's limit of feasible soil conductivities. Nevertheless, Patton (1986) and Ellington (1991; Ellington and Ferguson, 1991) asked an open-minded question: if infiltration were to be implemented in this region to meet the same standards that are met by detention systems, how much would it actually cost, in terms of land and money, compared with the conveyances and detention basins used in prevailing drainage practices?

The researchers designed hypothetical infiltration systems to replace existing detention systems on two urban sites in the Atlanta area as described earlier in this chapter. The sites were in different counties and thus under different runoff control standards, held different land uses and had been laid out by different design practitioners. On the 6-ha, gently sloping industrial site (Figure 5.1) a single detention basin was located at the major discharge point. On the 11-ha, steeply sloping residential site three detention basins were located in series to control discharge at one point, and one other basin was placed to control discharge at another point. On paper, the researchers lifted out the detention systems and replaced them with hypothetical infiltration systems designed to the same hydrologic standards. All other aspects of the land use—buildings, roads, parking lots and general earthwork—were unchanged, leaving a direct, controlled comparison between infiltration and detention.

In the first study, Patton (1986) based detention and infiltration sizing on the rational formula. Most of the infiltration basins were constructed with stone under parking pavements, unlike the existing detention basins which were in surface drainage swales. Basins located in parking bays were surfaced with open-celled concrete pavers to let surface runoff into the basins. Supplemental surface basins were located in planted areas.

Ellington (1991; Ellington and Ferguson 1991) repeated Patton's experiment using the Minnesota approach, applying the long-term water balance to background flows and the SCS method to design storms. In Ellington's designs all infiltration basins were located underground, with open-celled concrete pavers in one alternative and, in another, with impermeable asphalt paving and a grated drop inlet to let surface runoff into the stone and to allow maintenance access.

The cost results in both studies were similar. Costs estimated for Ellington's alternative systems are listed in Table 5.3.

In the table, basin area is the proportion of the development site dedicated exclusively to single-purpose stormwater basins. Considering the high cost of urban land, it can be economically beneficial not to dedicate large amounts of land exclusively to stormwater basins. In Ellington's hypothetical designs, infiltration was more favorable than detention: infiltration basins were constructed mostly underground, with the surfaces over them reclaimed for parking or planting.

Basin hydraulic capacity and total volume are indicators of basin construction costs associated with quantities of earthwork and construction materials, independent of specific unit costs at any moment in time. The total volume of a stone-filled basin

Table 5.3 Cost indicators for alternative drainage system on two sites in the Atlanta area (Ellington, 1991; Ellington and Ferguson, 1991).

Stormwater management	Area exclusively in basin (% of site)	Hydraulic capacity (m^3)	Total basin volume (m^3)	Construction cost ($)
		Residential site		
Detention	1.0	1,140	1,140	215,433.
Infiltration	0.1	913	2,280	128,582. with porous pavers, 89,441. with drop inlets
		Industrial site		
Detention	3.7	1,320	1,320	121,146.
Infiltration	0.0	1,630	4,070	255,968. with porous pavers, 103,062. with drop inlets

with 40 percent void space is 2.5 times its hydraulic capacity. At the industrial site the infiltration capacity was allocated partly to holding monthly background accumulations indicated by the long-term water balance, requiring larger capacity and total volume than detention. At the residential site the soil's higher hydraulic conductivity and areas available for large basin floors combined to eliminate background accumulations, making infiltration's capacity lower than detention's, although its total volume was greater.

The listed construction costs are based on 1989 unit costs. Costs for detention reflect inlets, culverts, basin excavation and basin outlet; the most expensive component was concrete culverts. Costs for infiltration reflect excavation, stone, filter fabric, and inlets or porous pavers; the most expensive component was stone fill or, where used, open-celled pavers. On the residential site infiltration was less expensive than detention whether constructed with porous pavers or drop inlets. On the industrial site infiltration was less expensive when constructed with inlets rather than open-celled pavers.

Thus, despite the cost of stone for underground infiltration basins, carefully planned infiltration systems need not cost more than equivalently designed detention systems. Where closely spaced infiltration surfaces and basins eliminate runoff near its source, and where each is sized to hold the design storm as well as all background flows, the expense of culverts to convey runoff downstream is eliminated. When these favorable findings in a seemingly unfavorable environment are combined with the decades of acceptable experience in California and on Long Island, it is clear that infiltration deserves to be tried in many regions where it is not yet considered conventional practice.

Determinants of feasibility

Tests of infiltration's feasibility for specific sites need to be as objective and controlled as those of Patton and Ellington. Following such a test, infiltration may or not be found feasible for a specific site, depending on site configuration, soil type, topography, land use intensity, etc.

Infiltration is most feasible when planned at the earliest stages of site layout. When site planning is at its best, provisions for infiltration merge with those of buildings, pavements, road and path alignments, grading and vegetation. To speed evaluation of early planning alternatives, rough regional indices of infiltration basin requirements can be learned from experience, in terms such as the floor area or hydraulic capacity required for each hectare of drainage area.

The comparison of cost in drainage systems is not infiltration vs. nothing; it is infiltration vs. the kind of system one would use if not infiltration—that is, some combination of conveyance and detention. In evaluating infiltration's cost, it should not be considered an add-on to a conventional runoff conveyance system. Culverts can be eliminated by planning infiltration from the beginning, so that only one primary drainage system—the infiltration one—is paid for, without building two systems side by side. Infiltration surfaces can replace impermeable pavements, eliminating runoff-producing areas. Infiltration basins can replace headwater culverts on a one-to-one basis: wherever there is a drainage inlet, curb cut or headwall that would lead to a culvert for conveyance, it can be replaced with an infiltration basin. Thus there can be a basin at every low point in a road profile, at every street intersection, at every crossing of a driveway over a roadside swale, at intermediate points along a street grade, and perhaps at every downspout discharge.

A costly aspect of infiltration on some sites could be a considerable area of land required to infiltrate the runoff from a given drainage area. The need to minimize the size of infiltration basins, through imaginative site-specific layout, construction and hydrologic design, is a real one.

Construction of infiltration basins need not require specialized contractors. Only local earth movers, pipe and stone suppliers, and landscape contractors are required. The key construction steps to make infiltration work are to check that volumes and elevations installed by a given contractor are adequate under design specifications, that the soil is not compacted, and that the capacity of the basin is not preempted by sedimentation.

Where infiltration cannot feasibly capture enough of the storm volume to control downstream peaks to the required degree, it can still be used to treat low flows for water quality. In this case, to control downstream flood peaks, an infiltration basin can be supplemented by surface detention of flow volumes that overflow or bypass the infiltration basin.

REFERENCES

Bianchi, W.C., and Dean C. Muckel, 1970, *Ground-Water Recharge Hydrology*, ARS 41-161, Washington, D.C.: U.S. Agricultural Research Service.

Chang, A.C., and A.L. Page, 1985, Soil Deposition of Trace Metals during Groundwater Recharge Using Surface Spreading, pages 609-626 of *Artificial Recharge of Groundwater*, Takashi Asano, editor, Boston: Butterworths.

Ellington, M. Morgan, 1991, *Comparison of Infiltration and Detention in the Georgia Piedmont Using Recent Hydrologic Models*, MLA Thesis, Athens: University of Georgia.

Ellington, M. Morgan, and Bruce K. Ferguson, 1991, Comparison of Infiltration and Detention in the Georgia Piedmont Using Recent Hydrologic Models, pages 213-216 of *Proceedings of 1991 Georgia Water Resources Conference*, Kathryn J. Hatcher, editor, Athens: University of Georgia Institute of Natural Resources.

Ferguson, Bruce K., 1990, Role of the Long-Term Water Balance in Management of Stormwater Infiltration, *Journal of Environmental Management* vol. 30, pages 221-233.

Ferguson, Bruce K., and Tamas Deak, (in press), The Role of Storm Flow Volume in Urban Drainage Problems, accepted for publication in *Water Resources Planning and Management.*

Ferguson, Bruce K., M. Morgan Ellington and P. Rexford Gonnsen, 1991, Evaluation and Control of the Long-Term Water Balance on an Urban Development Site, pages 217-220 of *Proceedings of 1991 Georgia Water Resources Conference*, Kathryn J. Hatcher, editor, Athens: University of Georgia Institute of Natural Resources.

Gburek, William J., and James B. Urban, 1980, Storm Water Detention and Groundwater Recharge Using Porous Asphalt—Initial Results, pages 89-97 of *International Symposium on Urban Storm Runoff*, Lexington: University of Kentucky.

Gerba, Charles P., and Sagar M. Goyal, 1985, Pathogen Removal from Wastewater during Groundwater Recharge, pages 283-318 of *Artificial Recharge of Groundwater*, Takashi Asano, editor, Boston: Butterworths.

Jackura, Kenneth A., 1980, *Infiltration Drainage of Highway Surface Water*, FHWA/CA/TL-80/04, Washington, D.C.: Federal Highway Administration and Sacramento: California Department of Transportation.

Ku, Henry F.H., and Dale L. Simmons, 1986, *Effect of Urban Stormwater Runoff on Ground Water Beneath Recharge Basins on Long Island, New York*, Water-Resources Investigation Report 85-4088, Syosset, New York: U.S. Geological Survey.

Leonard, R.A., G.W. Bailey and R.R. Swank, Jr., 1976, Transport, Detoxification, Fate, and Effects of Pesticides in Soil and Water Environments, pages 48-78 of *Land Application of Waste Materials*, Ankeny: Soil Conservation Society of America.

Maryland Water Resources Administration, 1984, *Standards and Specifications for Infiltration Practices*, Annapolis: Maryland Department of Natural Resources, Water Resources Administration, Stormwater Management Division.

Maryland Water Resources Administration, 1986, *Minimum Water Quality Objectives and Planning Guidelines for Infiltration Practices*, Annapolis: Maryland Department of Natural Resources, Water Resources Administration, Stormwater Management Division.

McCarty, Perry L., Bruce E. Rittmann and Martin Reinhard, 1985, Processes Affecting the Movement and Fate of Trace Organics in the Subsurface Environment, pages 627-646 of *Artificial Recharge of Groundwater*, Takashi Asano, editor, Boston: Butterworths.

McCuen, Richard H., 1974, A Regional Approach to Urban Storm Water Detention, *Geophysical Research Letters* vol. 1, no. 7, pages 321-322.

McCuen, Richard H., and Glenn E. Moglen, 1988, Multicriterion Stormwater Management Methods, *Journal of Water Resources Planning and Management* vol. 114, no. 4, pages 414-431.

Nellor, Margaret H., Rodger B. Baird and John R. Smyth, 1985, Health Aspects of Groundwater Recharge, pages 329-356 of *Artificial Recharge of Groundwater*, Takashi Asano, editor, Boston: Butterworths.

Nightingale, Harry I., 1978, Lead, Zinc, and Copper in Soils of Urban Storm-Runoff Retention Basins, *Journal of the American Water Works Association* vol. 87, no. 8.

Nightingale, Harry I., 1987a, Water Quality Beneath Urban Runoff Management Basins, *Water Resources Bulletin* vol. 23, no. 2, pages 197-205.

Nightingale, Harry I., 1987b, Accumulation of As, Ni, Cu, and Pb in Retention and Recharge Basins from Urban Runoff, *Water Resources Bulletin* vol. 23, no. 4, pages 663-672.

Nightingale, Harry I., 1987c, Organic Pollutants in Soils of Retention and Recharge Basins Receiving Urban Runoff Water, *Soil Science* vol. 144, no. 5, pages 373-382.

O'Hare, Margaret P., Deborah M. Fairchild, Paris A. Hajali and Larry W. Canter, 1986, *Artificial Recharge of Ground Water*, Chelsea: Lewis.

Patton, Stacy R., 1986, *Infiltration Basins as a Stormwater Management Alternative for the Piedmont Region of Georgia*, MLA thesis, Athens: University of Georgia.

Schueler, Thomas R., 1987, *Controlling Urban Runoff: A Practical Manual for Planning and Designing Urban BMPs*, Washington, D.C.: Metropolitan Washington Council of Governments.

Schueler, Thomas R., Peter A. Kumble and Maureen A. Heraty, 1992, *A Current Assessment of Urban Best Management Practices, Techniques for Reducing Non-Point Source Pollution in the Coastal Zone*, Washington, D.C.: Metropolitan Washington Council of Governments.

Washington State Department of Ecology, 1992, *Stormwater Management Manual for the Puget Sound Basin (The Technical Manual)*, Olympia: Washington State Department of Ecology.

Watschke, Thomas L., and Ralph O. Mumma, 1989, *The Effect of Nutrients and Pesticides Applied to Turf on the Quality of Runoff and Percolating Water*, Final Report for U.S. Geological Survey, University Park: Pennsylvania State University Environmental Resources Research Institute.

Weaver, Robert J., 1969, *Recharge Basins for Disposal of Highway Storm Drainage: Theory, Design Procedure, and Recommended Engineering Practices*, Research Report 69-2, Albany: New York State Department of Transportation, Engineering Research and Development Bureau.

Wigington, P.J., Jr., C.W. Randall and T.J. Grizzard, 1983, Accumulation of Selected Trace Metals in Soils of Urban Runoff Detention Basins, *Water Resources Bulletin* vol. 19, no. 5, pages 709-718.

Wittgren, Hans B., and Karin Sundblad, 1990, Removal of Wastewater Nitrogen in an Infiltration Wetland with *Glyceria maxima*, pages 85-97 of *Constructed Wetlands in Water Pollution Control*, P.F. Cooper and B.C. Findlater, editors, Oxford: Pergamon Press.

6

Soil Maintenance for Infiltration

The long-term performance of an infiltration surface or basin depends on the changes that occur in it while it is in operation and on how maintenance procedures respond to those changes. Infiltration surfaces and basins trap and therefore accumulate sediment and other pollutants. They increase the volume of water that passes through the soil profile, inducing changes in soil structure and biotic growth. A need for maintenance is inevitable in an effective pollutant trap and dynamic soil surface.

A maintenance program for infiltration need not necessarily be more expensive than that for any other stormwater management approach, but it must always be anticipated. Care in design and construction is required to reduce future maintenance demands and to allow maintenance, when it is required, to be rapid, economical, convenient and effective.

Table 6.1 Conditions in infiltration and detention basins in three counties in Maryland (data from Lindsey, Roberts and Page, 1992a; data for subsurface infiltration basins are the total for "infiltration trenches" and "dry wells").

Condition	Dry Detention	Wet Detention	Total Detention	Surface Infiltration	Subsurface Infiltration	Total Infiltration
	Structural condition (% of basins)					
Structural failures	17	10	16	0	4	3
Erosion at inlet or outlet	29	38	31	43	5	13
Bank erosion	53	48	52	79	16	30
	Flow & ponding (% of basins)					
Inappropriate ponding	36	17	32	50	20	26
Clogging of facility	28	14	25	43	29	31
Water bypassing facility	9	0	7	14	24	21
	Sediment (% of basins)					
Sediment inflow	53	55	53	64	55	56
Excessive sediment	51	41	49	57	50	51
Inlet or outlet clogged	27	10	24	0	18	13
Significant volume lost	20	21	20	36	na	na

That infiltration surfaces and basins undergo changes that require maintenance responses was illustrated by a survey of basins in Maryland (Lindsey, Roberts and Page, 1992a), some results of which are summarized in Table 6.1. Infiltration basins are susceptible to inadvertent ponding, because there is no surface outlet. On the other hand, infiltration does not share detention's problems of outlet clogging (because there is no principal outlet other than the soil) and structural failure (structural dams and conveyances are seldom part of infiltration basins). Both infiltration and detention can experience excessive sediment and debris, internal erosion, and loss of storage volume by sedimentation.

SOIL DEVELOPMENT DURING INFILTRATION

In an operating infiltration surface or basin, events of precipitation, runoff inflow, ponding and infiltration are repeated and in some cases prolonged. In this environment distinctive and important physical and biological processes are active at and near the soil surface.

Physical processes

Sediment mobilization

Infiltration surfaces and basins accumulate inflowing sediment. Sediment can be transported from their watersheds and laminated where it stops. The potential to develop an impeding surface layer is present in any closed basin or vertically drained surface.

Suspended sediment includes any undissolved solid matter. Natural soil constituents—sand, silt, clay, organic matter and, in arid areas, fine particles of calcite—can be delivered with inflowing runoff, depending on the intensity of rainfall, the types of soils, topography, vegetation, land use, and the exposure and disturbance of watershed soil by construction or earth-moving.

In urban areas many other materials can be added. Accumulated substances observed on the floors of Long Island basins included well graded mixtures of fine sand, silt and clay; natural litter such as grass clippings, leaves and twigs from unpaved areas; and organic matter such as asphalt, grease, oil, tar, and rubber particles from roads and parking lots (Aronson and Seaburn, 1974).

Weaver (1969, p. 50-51) characterized in detail the material deposited in one basin near the Long Island Expressway. The mineral portion was a well graded mixture ranging from 4 mm down to 2 percent finer than 0.002 mm. Organic content was 5.5 percent of dry soil weight, which is high enough ordinarily to give soil a dark color, but this sediment was only light brown. Although surface-active clay minerals were absent, upon drying the sediment shrank appreciably, the surface mat cracked, and the pieces curled up into saucer shapes. Microscopic examination revealed an astonishing quantity of minute fragments of undecomposed paper, cardboard, leaves and grass that gave the material its organic content and light color. When wetted, these formed a thin, sticky matrix that gave the material its cohesion. Bits of metal, both magnetic and non-magnetic, were also observed. Thus the basin was a repository for the products of wear and waste from urban land use and highway travel, in addition to the products of soil erosion.

Another source of mobile suspended matter in an infiltration basin is the breakdown of soil structure on the basin floor. This may be the most important source of sediment and reduced conductivity in a surprisingly large proportion of basins. Soil breakdown can occur in almost any soil that is not protected from rainfall by vegetation or mulch. The energy of raindrops throws particles about the soil surface, beating down soil structure. The extent of the sealing is a function of the energy applied and the aggregate size (Jennings, Jarrett and Hoover, 1988).

Some soils break down so readily that they break down in the presence of water without the energetic disturbance of rainfall. These soils are dispersive because of a zone of polarized water molecules around individual clay particles. The thickness of this zone depends on the charge of the cations bound to the clay surfaces. Generally, monovalent cations such as Na+ and K+ disperse soils because of their high degree of hydration, leading to structural collapse and low hydraulic conductivity. Divalent cations with low hydration such as Ca++ and Mg++ tend to flocculate soils, leading to structural aggregation and increased conductivity.

Another potential source of sediment is fluvial erosion inside an infiltration basin, where inflowing runoff discharges at high velocity onto the basin floor or down a side slope. The occurrence of this type of erosion depends on the construction of the basin and its inlets.

Physical sealing

Suspended particles eventually lodge in pores between larger particles. This in turn traps still smaller particles. Inter-aggregate cavities are diminished. Macropores can be partly or completely filled. Clay lamellae can eventually join to form a layer of low conductivity. In general the coarser the underlying soil and the finer the solids in the water, the more severe the relative clogging (Behnke, 1969). Even low concentrations of suspended matter, when prolonged, can gradually reduce the conductivity of a soil not protected by vegetation. Once a crust is established, conductivity does not continue to decline with further deposition and crust thickness (Behnke, 1969).

Surface crusts formed by in-place disaggregation and reworking of soil material ("structural crusts") can form in almost any unprotected soil except the coarsest sands and gravels (Shainberg, 1992). The crustal layers are typically less than 2 mm thick. They are characterized by greater density, finer pores, and lower conductivities than the underlying soil. In the field they typically have a slick appearance and contrast sharply with the underlying porous, open soil. They are formed importantly by the physical impact of falling raindrops mechanically breaking down aggregates and compacting the soil surface. Physicochemically dispersed clay particles migrate into the soil with the infiltrating water and clog the pores immediately beneath the surface. The susceptibility of soils to structural crusting depends on clay content, mineralogy, organic matter content, and chemical constituents in the soil water.

Another type of crust is a depositional or sedimentary crust, formed by the deposition of sediment brought to the soil surface by flowing water (Shainberg, 1992). Dispersed clay particles can settle from turbid water in an orientation parallel to the soil surface, forming an effective seal. In some cases sediment deposition is associated with a visible alluvial cone where runoff discharges into the basin.

In some circumstances mobile suspended particles can migrate far below the soil surface. Fine-grained sediment can travel with infiltrating water several centimeters into a coarser-grained soil matrix (Weaver, 1969, p. 51, 54-55). Both coarse and fine particles can move downward via macropores. In a Texas playa basin underlain by calcareous clay loam with many large continuous pores, 50 percent of the sediment entering the basin moved deeper than 45 cm into the soil (Goss, Smith, Stewart and Jones, 1973). Very fine grains of montmorillonite can be carried further into the soil, even more than 10 m (O'Hare et al., 1986, p. 10).

On an uneven basin floor, suspended solids accumulate more in topographic depressions than on side slopes of ridges or humps. Only a few centimeters in relief have resulted in differential deposition (Jones, Goss and Schneider, 1974; Nightingale and Bianchi, 1977). Thus crusts may not be continuous in some basins.

Flaws in the crustal layer allow infiltrating water to bypass the crust (Moss and Watson, 1991). Flaws occupying only 10 percent of the soil area can raise the infiltration rate to that of uncrusted soil. Apparently water that penetrates through the flaws infiltrates three-dimensionally into the underlying uncrusted soil.

Mulch and vegetative cover can have an effect opposite to crusting, increasing soil conductivity by maintaining structure, porosity and crustal flaws. A cover of mulch, vegetation and plant litter intercepts raindrops before they strike the ground, mitigating their disturbing and sealing effect. Growing vegetation and associated fauna build macropores.

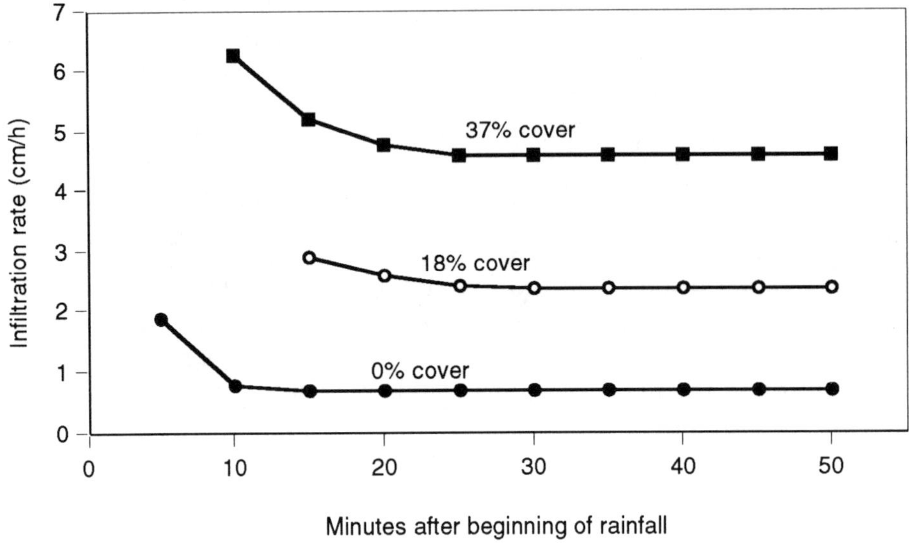

Figure 6.1 Infiltration rates into clay loam in New Mexico with different densities of vegetative cover (after Smith and Leopold, 1942).

Figure 6.1 shows infiltration rates on experimental plots in New Mexico with different coverages of grass and desert shrubs. Between bare soil and a coverage of 18 percent, the final infiltration rate increased from 0.69 cm/h to 2.41 cm/h, and then doubled as the vegetative cover doubled, to 4.75 cm/h. Thus coverages even of less than 50 percent can greatly increase surface conductivities.

Figure 6.2 shows a similar effect from nonliving mulch, and emphasizes the importance of the immediate soil surface in crusting and its prevention. Duley (1939) sprinkled sandy loam continuously at a rate in excess of intake, initially with a straw mulch, yielding a nearly constant final infiltration rate of 3.05 cm/h. When he removed the straw and exposed the soil to the falling drops of water, a sealing layer 1 mm thick formed, reducing the infiltration rate to one fifth the rate of the mulched soil. He then removed the surface crust and covered the soil with burlap. The infiltration rate immediately rose to a constant rate of 4.09 cm/h, showing that the soil beneath the thin, compact surface layer was still in condition to permit rapid penetration of water, even though it was thoroughly wet. When he removed the burlap, the infiltration rate again dropped to its previous low level, controlled by a new surface seal.

Humus also has an effect opposite to crusting, maintaining structure and porosity. The organic matter produced by vegetation stabilizes soil aggregates. Even fine-grained soils can have high infiltration rates if there is a large amount of decomposed organic material in the surface horizon.

Microbial processes

Large numbers of bacteria and numerous species of algae, actinomyces, yeasts, molds, fungi and protozoans are present in surface soils. Environmental conditions such as sunlight, temperature, oxygen supply, pH, salinity and nutrient supplies influence the number, kinds and biochemical activities of microbes (Zobell, 1975).

Under prolonged ponding, types of microbial growth occur that can cause or compound soil clogging. Prolonged ponding isolates the soil ecosystem from the atmosphere, changing the system from aerobic and oxidizing to anaerobic and reducing

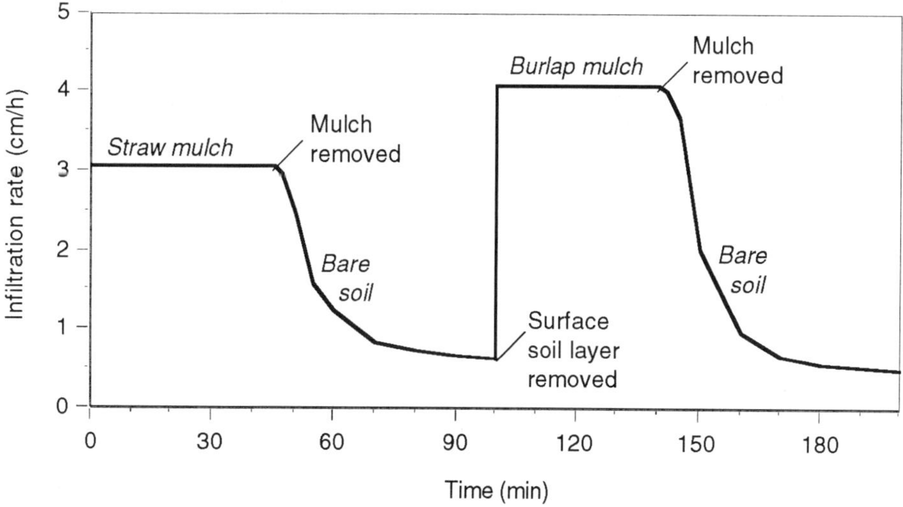

Figure 6.2 Infiltration rate of a sandy loam under continuous sprinkling of water at a rate in excess of intake, with a series of four surface conditions (after Duley, 1939, Figure 9).

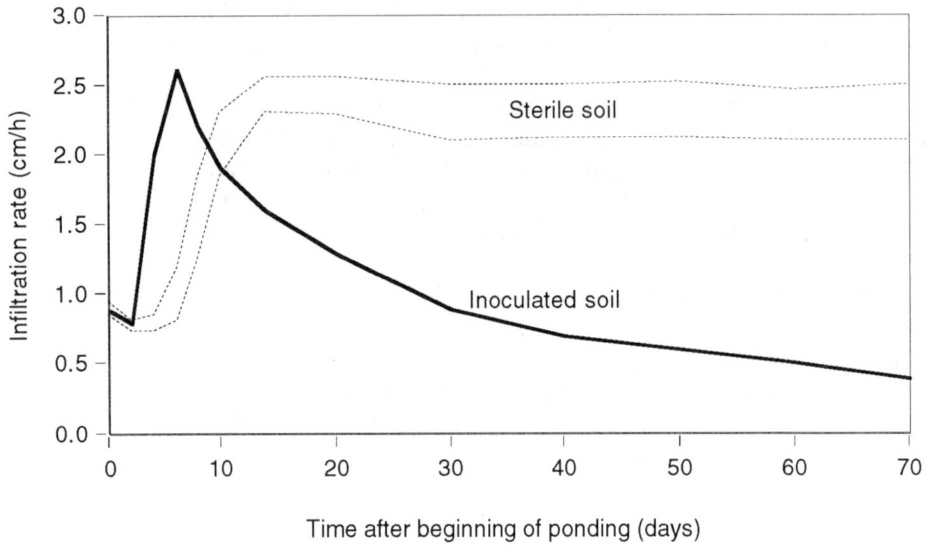

Figure 6.3 Infiltration through the Hanford loam of southern California during prolonged ponding (after Allison, 1947).

(Faulkner and Richardson, 1989). Prolonged ponding can be initiated by a basin floor constructed with insufficient area to infiltrate all inflow quickly, or by physical sealing. Similar anaerobic conditions might develop in the enclosed void space of some subsurface basins.

Under prolonged (more than 5 or 10 days) submergence of the soil, relatively consistent trends in infiltration rates have been observed (Allison, 1947). The pattern is illustrated by the "inoculated soil" curve in Figure 6.3, which shows the result of a laboratory measurement on the Hanford loam of southern California. In the first stage of ponding (approximately the first day), infiltration rate declined from its initial level, perhaps because of swelling and dispersion of the soil, leaching of salts, and reduction of matric suction by infiltrated water. In the second stage (lasting up to several days), infiltration rate increased to higher than its initial level, due to the dissolving and removal of air trapped in the soil pores, freeing pore space for percolation. In the third stage, lasting as long as the ponding continued, the infiltration rate declined slowly, approaching a very low level, due to clogging of soil pores with microbial products. The other curves in the figure show the upper and lower limits of infiltration rate under sterile conditions, free from microbiotic clogging.

A large quantity of decomposable organic matter accumulated in anoxic soil assists biological clogging. At the large Leaky Acres facility, during prolonged ponding, organic matter accumulated near basin margins where floating organic debris was driven by wind and waves. Poor soil drainage made biological clogging evident where the organic carbon content of the soil exceeded about 5 percent by weight (Nightingale and Bianchi, 1977).

Organic matter trapped in anoxic conditions does not contribute to soil aggregation and porosity as it does in aerated soils. On the contrary, it may reduce soil conductivity. In the presence of oxygen, decomposition is complete and produces a stable humus. In the absence of oxygen, organic decomposition is incomplete, leaving a slimy intermediate product. Clogging may be due to the accumulation of polysaccharides or to the sealing of soil pores by microbial cells and the accumulation of carbon dioxide gas.

In addition, prolonged saturation inhibits most types of animal and root activity that could form macropores and counteract soil sealing (Beven and Germann, 1982).

Experience with infiltration basins and land application of wastewater indicates that progressive blockage of soil is not generally a problem where the soil is "rested" between wettings (Driscoll et al., 1986, p. 24). For infiltration basins with dry or occasionally ephemeral regimes, the intermittent nature of ponding provides such rest periods.

Observed results

The results of three monitoring programs illustrate the interactions of developmental processes in the field.

On Long Island, Aronson and Seaburn (1974) associated prolonged ponding in basins with the types of land use in the watersheds, and thus with types of inflowing sediment. In residential areas, the detritus deposited on basin floors was relatively porous and friable in nature, facilitating infiltration such that only 7 percent of the observed basins contained water between storms. In commercial and industrial areas, 28 percent were water-containing, probably as a result of large influxes of asphalt, grease, oil, tar and rubber particles from large impervious parking areas. About 9 percent of basins draining highways were water-containing; this number may have been held down by relatively regular maintenance by public highway agencies and relatively frequent use of wells in the basin floors.

Also on Long Island, Seaburn and Aronson (1974) noticed low infiltration rates in the Deer Park basin compared to two other basins in the area. Average infiltration rate in the Deer Park basin was only 22 to 25 percent of that in the other basins. Ponding at Deer Park lasted 2 to 3 days after a storm, but only a few hours at the other two. Deer Park's infiltration rate had declined following the onset of residential construction in the watershed and the visible arrival of sediment in runoff. The other basins were in stable watersheds and had abundant vegetation keeping the floor surface soil loose and permeable. Seaburn and Aronson attributed Deer Park's low infiltration rates to clogging by silt, clay and organic debris washed in from the construction sites, and to lack of protective plant growth on the basin floor. Perhaps vegetation could become established at Deer Park with stabilization of the watershed. However, the Deer Park basin occupied a smaller portion of its drainage area than those to which it was compared—1.2 percent as opposed to 3.3 to 3.5 percent—and the concentration of runoff on a smaller floor area may have contributed to a wetter hydrologic regime.

In North Carolina, Chescheir, Fipps and Skaggs (1988) observed the floor materials in two infiltration basins on coastal islands. At both sites the ambient soils were sandy and permeable. However, the surface soils in the basin floors were finer than either the basins' subsoils or the average soils in the basins' drainage areas. The researchers attributed the fine surface texture to washing off of impervious surfaces in the drainage areas, but it could have resulted at least partly from in-place crust formation. The vertical conductivities of the floor sediments (the top 7.5 cm) were between 2.5 and 7.2 percent of that of the nearby soil outside the basins. At Surf City on Topsail Island, the basin occupied only 1.6 percent of its drainage area and held standing, fluctuating water most of the time. The water table was only 20 cm below the basin floor. The fine-textured surface layer was between 3 and 5 cm thick and contained a dense root mass from growing vegetation. The other basin, on Bald Head Island, had been constructed to meet a volume requirement similar to that at Surf City, but the basin occupied 18 percent of its drainage area. The basin regime was dry. Soils were particularly sandy and excessively drained. The water table was a meter or more deep. Vegetation consisted only of sparse dune grasses. The fine-textured surface layer was less than 5 mm thick and did not contain many roots. The conductivities of the two surface layers were similar despite the difference in vegetative density, perhaps because prolonged saturated and anaerobic conditions at Surf City contributed to microbial clogging.

CONSTRUCTION TO COMBAT SEDIMENT INFLOW

An imperative aspect of the design and construction of infiltration surfaces and basins is controlling the inflow of suspended sediment. Wherever infiltration is to be practiced it must be accompanied by a vigorous erosion and sediment control program during and after construction. However, no at-source sediment control should be assumed perfectly effective.

Sediment traps of various kinds have been built upstream of some basins. Trap design must follow the principles of sediment settling and filtering (Roesner, Urbonas and Sonnen, 1989). If sediment is to be captured and held, then a specific volume must be left in the trap for the quantity of sediment that is anticipated during the life of the structure or until sediment removal can be expected. Installation of any sediment trap must be accompanied by a realistic anticipation of the sediment removal requirement.

Coarse sediment can settle out in still water. Large settling basins with long residence times remove sediment effectively and can provide preinfiltration storage during large runoff events. Small drop inlets used with small wells in California have been found to trap only the heaviest dirt and trash, allowing finer suspended matter to flow into the well (Jackura, 1980, p. 25).

On Long Island, settling basins like that in Figure 6.4 have been constructed as a way to reuse infiltration basins that large accumulations of sediment and debris have permanently clogged, or where continual inflows of sediment and debris have made repeated rehabilitation too costly (Aronson and Seaburn, 1974). Most of these basins drain large parking areas and heavily traveled roadways that contribute high concentrations of asphalt, grease, oil, tar, and rubber particles. They are used to remove sediment and debris from runoff. The relatively clean overflow is discharged, by way of pipes or flumes, to another nearby basin for infiltration.

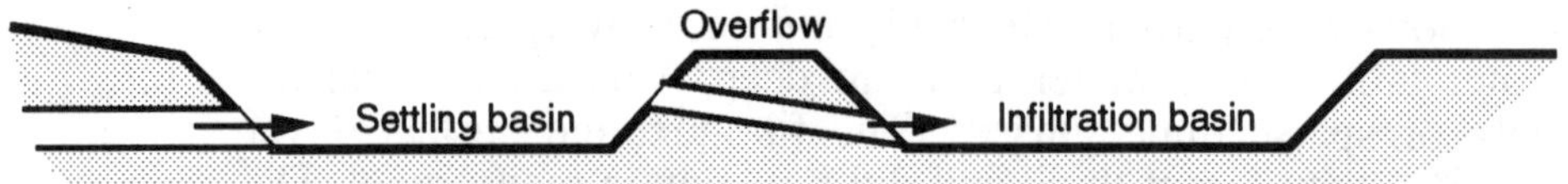

Figure 6.4 Settling basin on Long Island (after Aronson and Seaburn, 1974).

Within infiltration basins on Long Island, settling areas like that in Figure 6.5 have been constructed by grading the floors in two levels (Aronson and Seaburn, 1974). The lower level is intended to collect trash and sediment washed in by runoff. The upper level, about 0.5 m higher, is an auxiliary infiltrating area that receives overflow from the flooded lower level. Settling areas can be expected to have wetter regimes than infiltration terraces and could be designed as ephemeral wetlands. The upper level can be expected to be relatively better drained, vegetated with mesic species, and maintained by a more open soil structure and slower accumulation of sediment.

Clay and colloids can stay suspended for very long times, eventually being carried through settling ponds into infiltration areas. The electrochemical charges on their surfaces can keep them suspended away from each other, like the north poles of two magnets. Ponded water remaining muddy or clouded by sediment in still conditions long after storms is a sign that dispersive particles are present.

Adding a flocculant such as gypsum or alum to water can reverse the effect of dispersed particles' surface charges, encouraging aggregation and settling of large, heavy aggregate particles (Weaver, 1969, p. 51, 54-55). Samples of inflow water can be analyzed to determine the quantity of flocculant required. Powdered gypsum can be added to a pond as a single treatment. However, the flocculating effect will continue only as long as fresh flocculant continues to be added to replace that leached out of the system. For a longer treatment, a "riffle" of gypsum rock can be added to a channel just upstream from the basin, at an elevation above the channel bottom that will be reached only by significant storm flows. Alternatively, an automatic proportional dispenser at the basin inlet, controlled by a float or vane, can add a liquid solution of flocculant proportionally over a range of flow rates.

Appropriately selected filters can effectively remove all but the finest sediment from throughflowing water. Ferguson and Gonnsen (1993) used filter fabrics with gabions, rock check dams and perforated standpipes, causing sediment to collect in adjacent basins. Sand filters have been attempted in Florida and Delaware. No filter can separate from water the finest colloidal particles, the dimensions of which approach those of water molecules.

Gently sloping, densely vegetated "buffers" can remove sediment from slowly moving sheet flow. As the sediment accumulates, the average elevation of the buffer area rises (Wilson, 1967). Grass has served as a good filter material in California (Jackura, 1980, p. 72). If sediment-laden water is allowed to trickle down a densely turfed slope, most of the suspended material is left behind within a few meters of surface travel.

Where the full volume of runoff from a site need not be infiltrated, the sources of runoff within the site can be selected for inflow to infiltration basins based on their relative freedom from sediment (the quality of the remaining water is left to be treated by surface means such as artificial wetlands). The best conditions in two surveys of basins in Maryland (Pensyl and Clement, 1987; Lindsey, Roberts and Page, 1992b) were observed in "dry wells," which are small stone-filled basins usually receiving only roof water through downspouts. The nearly sediment-free source of runoff may be the cause of the good results. Thus one concept for a relatively complete infiltration plan would include small dry wells receiving relatively clean runoff from roofs, and maximum use of permeable pavements to spread out contaminated runoff over the largest possible soil surface.

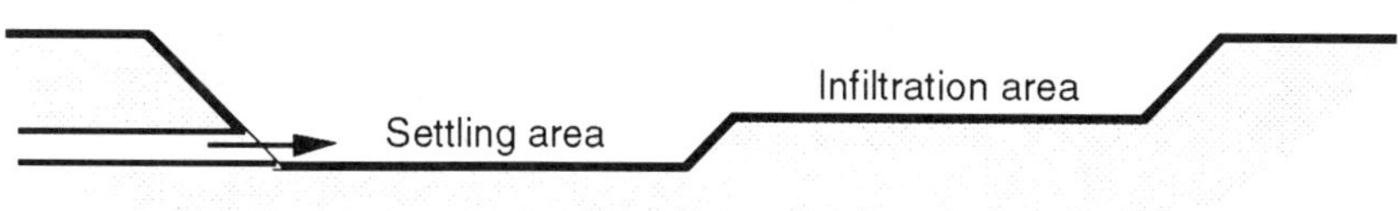

Figure 6.5 Settling area on Long Island (after Aronson and Seaburn, 1974).

To prevent erosion within a basin, dispersed runoff can be delivered to the basin down broad, very gentle, vegetated slopes. Where concentrated runoff flows toward a basin, it should be intercepted by swales or inlets and guided onto the basin floor at low velocity, possibly with energy dissipation at the discharge onto the floor (Weaver, 1969, p. 48).

Trucks must be prevented from tracking or spilling dirt onto a porous pavement. If the pores are plugged under pressure, they are not amenable to cleaning.

No sediment trap, filter or runoff source is perfect. It must be assumed that solids will reach the infiltration surface at some rate no matter how gradual. In addition, every surface and basin is susceptible to internal developmental processes such as slaking of the soil surface and growth of algae and bacteria. No matter how the inflow is treated, design and construction must aim to combat soil crusting, and access must be provided to the floor so it can be monitored and maintained.

CONSTRUCTION TO COMBAT SOIL CRUSTING

Even without significant sediment inflows, crustal formation by breakdown of in-place soil can greatly reduce porosity and conductivity. An infiltration surface or basin's hydrologic design, construction processes and plantings create the conditions in which soil development may or may not result in surface crusting. The surfaces and basins least vulnerable to declining infiltration rate are those with dry hydrologic regimes and dense vegetative covers.

Complete drying between storms is accomplished in basins with large enough floors in relation to the conductivity of the floor soil. Drying stimulates aerobic humus-building and soil aggregation and inhibits clogging by microbial growth. With an aerated root zone, vegetation can root deeply and maintain macropores. A dry hydrologic regime for a proposed basin can be confirmed by applying the long-term water balance and by evaluating ponding time following the design storm. In calculations, the conductivity safety factor Sf must be conservative (no higher than 0.5).

Almost any mulch or vegetation helps to disperse rainfall energy, maintain crustal flaws, build macropores and maintain total infiltration rates. Well-established Bermuda grass on a basin floor grows up through sediment deposits, forming a porous turf and preventing formation of an impermeable layer. Planting of Bermuda grass and, after a period of time, establishment of a dense growth and a mulch of old grass, have led to substantial improvements in infiltration rates in previously bare basin soils. Clumped or woody vegetation could maintain microtopography to stimulate differential accumulation of sediment and flaws in surface crusts. Mulch should probably be added after basin construction, tillage or any other soil disturbance for immediate soil protection until vegetation is established. If the mulch is made from natural materials, its decomposition can add organic matter to the soil.

Where the soil is unvegetated or recently disturbed, adding gypsum (calcium sulfate) to the soil surface can inhibit crusting by discouraging dispersion of clay particles (Shainberg, 1992). Gypsum amendments have been found at least somewhat effective on almost any type of soil. Figure 6.6 illustrates the effect of gypsum on crusting and infiltration rate of a sandy loam. Different samples of the soil were chemically given two levels of exchangeable sodium percentage (ESP), which is as-

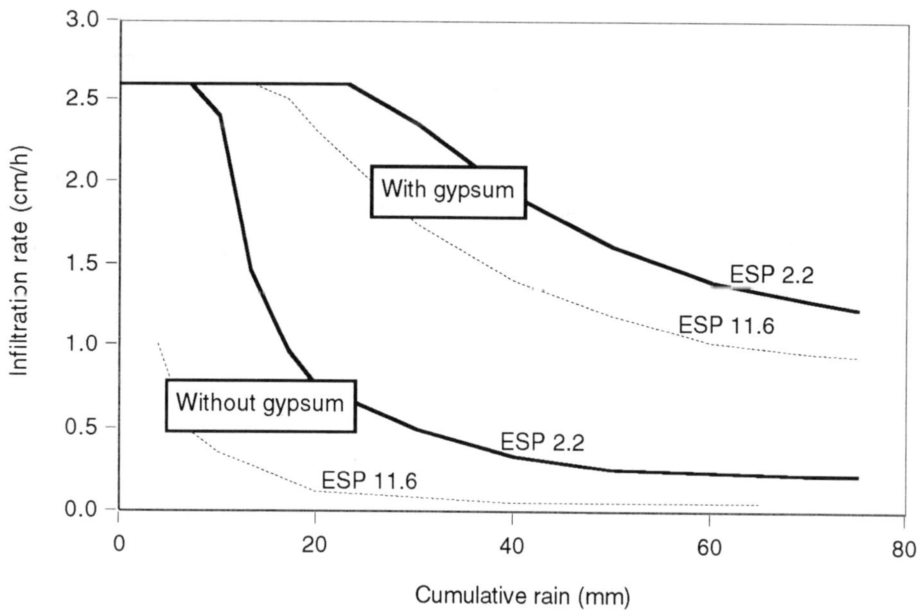

Figure 6.6 Infiltration rate of the Netanya sandy loam of Israel given two different soil chemistries (exchangeable sodium percentage, ESP), with and without gypsum amendment (after Kazman, Shainberg and Gal, 1983, Figure 2).

sociated with increasing chemical dispersion and therefore crusting under raindrop impact. When the unamended soil samples were sprinkled with water, a crust formed and infiltration rate declined quickly to a low level. When identical samples were amended with gypsum, the onset of crusting was delayed (indicated by the horizontal part of the curves on the upper left) and had less influence on infiltration, with infiltration rate declining slowly to a final level many times higher than that of the unamended soil.

Gypsum is available in powdered and pelletized forms from agricultural and garden suppliers. Some manufacturers mix gypsum pellets with organic materials and sell the mixture with trade names like "Clay Buster." Gypsum should probably be added just before mulching to almost any newly excavated basin or recently disturbed soil to improve and protect the soil during initial seedling establishment. The small amounts of gypsum needed at the soil surface can make such amendments highly cost effective. Gypsum is a naturally occurring salt with a self-limiting concentration in water, so it is not likely to become a pollutant even if over-applied. However, gypsum's effect is only temporary, as it is eventually leached out of the soil by infiltrating water. Where the soil is not to be vegetated for long-term protection against crusting, gypsum may have to be replenished routinely as part of the basin's maintenance program.

In the future, polyacrylamides (PAs) may be used to stabilize favorable soil structure (Shainberg, 1992). Polyacrylamides are a class of polymers derived from petroleum. Different members of the class have different effects on soils. Correctly selected PAs bind soil particles together. Experimental treatments combining gypsum and appropriately selected polymers have been highly effective in maintaining hydraulic conductivity of unvegetated soils in the laboratory. Immediately following addition of gypsum to establish structure, correctly selected PAs can be sprayed onto the soil surface, preserving the soil's structure even though the gypsum leaches out later. The PAs now being marketed for soil moisture retention are probably not suited to this purpose. Practical application of PAs in stabilizing soil structure awaits commercial selection and marketing of a PA specifically for this purpose and field experience combining PAs with different levels of vegetative cover.

CONSTRUCTION FOR MAINTENANCE ACCESS

The designation of responsibility for monitoring and maintenance of stormwater structures depends on the types and qualities of local resources. In recent years, public agencies have been increasingly selected over private landowners for this purpose. In some cases the agency is an existing one such as a public works department; in others it is a specially created stormwater utility supported by taxes or fees. Easements for maintenance access, connecting infiltration surfaces and basins to public rights of way, are essential.

Maintenance of a stormwater basin might be for hydraulic competence alone, or for aesthetics and multiple use. Aesthetics and human use tend to demand more frequent and exacting maintenance than does hydraulic competence. Making a basin into a positive amenity, with maintenance assigned to a park agency or other human-oriented organization, is a way to assure diligent monitoring and maintenance.

Subsurface basins require distinctive built-in components in order to be physically accessible for monitoring and for performance of specific maintenance procedures. Access to the infiltration surface allows monitoring and maintenance at the ultimate place where sediment would accumulate and water would stand. A well or inlet must be included, wide enough, perhaps half a meter or more, to lower hoses, pumps or other equipment into. A narrow "observation well," permitting only observation of sediment and water as they accumulate, is inadequate. Many local governments specify minimum diameters of drainage culverts to make them accessible and maintainable; the same minimum size could be construed as the minimum size of a well or inlet in a subsurface basin.

Using conventional-looking drainage inlets for access is appropriate for typical maintenance personnel. If a subsurface basin has overflowed to the point that water is standing on the surrounding ground, a drainage inlet is the most obvious place a typical crew would look for problems, whether or not they had been trained in the specific objectives or operation of infiltration basins. When accumulated mud is found in a drop inlet, pumping it out would be a typical crew's obvious response—and it would be the right one.

Where water is let into a subsurface basin through permeable pavement, an inlet may not seem necessary for letting water in, but it is still necessary for monitoring and maintenance access. Where an inaccessible subsurface basin has failed to the point that water has ponded on the pavement surface, the development of the problem would not have been seen until the basin had become entirely choked with sediment or filled with water. The only way to clean out an inaccessible subsurface basin is to tear it up and reconstruct it. Initial design and construction must bend over backward to prevent tearing up basins from being the only maintenance option, by providing maintenance access all the way to the infiltration surface on the basin floor.

MONITORING AND MAINTENANCE SCHEDULE

Incidental maintenance for small, simple problems can abolish the need for later drastic rehabilitation or reconstruction. A system for keeping thorough monitoring and maintenance records can help to identify the structures requiring special attention and to plan future work.

Ordinary infiltration surfaces and basins should be monitored on the same type of schedule as other drainage structures in the community. Different communities are accustomed to, and are organized to deal with, different kinds of maintenance schedules. In many communities maintenance is scheduled only as needed, in response to casual observations during everyday work or complaints from residents. The most obvious signal that an infiltration basin is not working properly is the same as that for any other drainage structure: prolonged ponding over the surface of the adjoining ground. Waiting until such problems occur before taking maintenance action allows more frequent overflows than preemptive monitoring and maintenance, but it allows a small labor force to concentrate its efforts upon the relatively few basins that actually need attention. Where hundreds or thousands of basins are in operation, monitoring and maintenance resources must be focused and disciplined; monitoring every basin once a year or more may be unfeasible.

If scheduled preventive monitoring is to be done, it should be done during low-flow periods, when basins are expected to be dry. Such periods occur after prolonged warm and dry weather, such as late summer and early fall in much of the continental United States. To predict months when background accumulations would be lowest during an average year, the long-term water balance can be calculated. Within the low-flow months specific inspections can be planned for periods following particularly dry, hot weather. If excess standing water is encountered at such a time, it is because the basin is draining slowly: the basin is not infiltrating as expected. For many basins, observation after wet periods or major storms is not likely to provide actionable data because, in wet periods, typical basins are expected to hold standing water; standing water only confirms that it has rained in the watershed recently. Standing water soon after a wet period indicates a definite problem only in those specialized basins where rapid drawdown is required for scheduled human use or other reasons.

If upon inspection of any basin during a low-flow period enough sediment or water is found to be accumulated that the basin's capacity is measurably reduced, then the existence of a problem has been recognized. However, the problem's duration, severity and cause have not yet been determined.

An accumulation of water or sediment may be the result of recent large storms or pulses of sediment from temporarily exposed construction sites. In these cases, only mild, discrete, short-term action is called for. Pumping or otherwise removing all sediment and standing water using manual or truck-mounted equipment can restore the capacity. If there are obvious signs of inflowing sediment from actively eroding construction sites or drainage channels, the sources must be immediately mitigated with silt fences, check dams and sediment traps, and quickly vegetated.

After simple and immediate restoration, further monitoring can determine whether the problem is chronic. Additional monitoring, including monitoring in wet periods, is now justified in order to determine the basin's frequency and depth of ponding, its permanent minimum water level if any, and its drawdown time following storms.

TREATMENT OF EXCESSIVELY PONDED BASINS

If excessive ponding in an infiltration basin is recurrent or chronic, then calm, objective study of the specific problem, its causes, and alternative responses is required. The basin's design, construction and condition, the condition of the watershed, and the performance of the basin in the context of the specific weather events that have occurred, must be reviewed sufficiently to determine conclusively the cause of the standing water or sediment. It should be a matter of religion that the cause of excessive ponding is to be conclusively determined before treatment for it is administered. Only responses that are firmly believed to address the known cause should be selected. The more costly or irreversible a proposed treatment, the more sacredly should this rule apply.

Causes and treatments of excessive ponding

Aronson and Seaburn (1974) distinguished two types of causes of excessive standing water in infiltration basins: flooding and clogging. Figure 6.7 shows the different principal directions of water movement from flooded and clogged basins. Flooded basins are those that contain water because they intersect a water table or were constructed with insufficient floor area to infiltrate inflows into preexisting soil of given hydraulic conductivity. Clogged basins are those that contain water because slowly conductive layers formed or were deposited in the basin after the basin was put into operation.

Basins that intercept the regional water table can be identified by examination of nearby pond elevations to identify regional water table gradients, by drilling to locate nearby water table elevations, or by interpretation of regional water table maps, and finding the difference in elevation between the basin floor and the regional water table. Of 194 excessively ponded basins on Long Island studied by Aronson and Seaburn (1974), 22 intersected the regional water table all or part of the year. Included in this total were four basins that had been constructed during a 1962-66 drought and that subsequently became flooded as ground water levels recovered after the drought. If a basin intercepts the regional water table, the wet regime of the basin can be accepted and accommodated through landscape design, and supplemental capacity can be acquired by lateral excavation.

Basins that intersect a perched water table can be identified only by obtaining detailed information about the lithology of the underlying materials with borings or other direct observations. A perching soil layer can be punctured with a supplemental well.

Basins with floor areas too small for the conductivity of the ambient soil can be identified by hydrologic modeling following investigation of the soil, the drainage area and the basin's dimensions. A floor area can be enlarged by excavating. If soil of higher conductivity is present deeper in the profile, supplemental wells can be constructed in the basin to puncture the

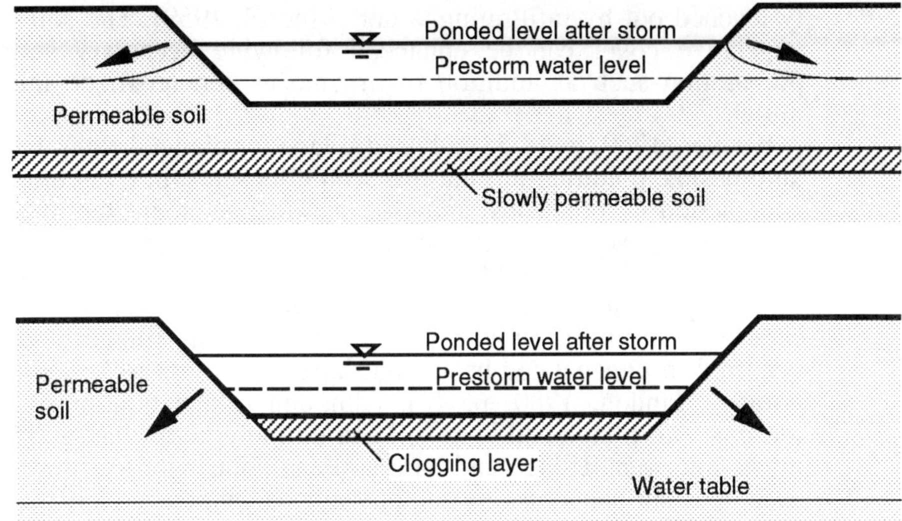

Figure 6.7 Flows from infiltration basins containing excessive standing water: above, a flooded basin; below, a clogged basin (after Aronson and Seaburn, 1974).

inhibiting layer. Runoff inflow can be reduced by adding small basins upstream, close to the runoff source. Additional basins can be constructed downstream and the existing basin reclaimed as a settling area.

Treatment of a clogging layer

Basins in which a crust—a clogging layer—has formed or been deposited can be identified by examining the thickness, composition, structure and hydraulic conductivity of the materials on and beneath the floor. A variety of responses can be appropriate.

The layer can be made more permeable. Pedestrian and other traffic, if any, can be limited to reduce compaction. Applications of flocculating agents such as gypsum can aggregate soil and increase infiltration rate (Muckel, 1959; Shainberg, 1992). However, increases in infiltration rate due to gypsum and related additives are only temporary; the chemicals are eventually leached out by infiltrating water (Muckel, 1959). The soil's vegetative cover can be enhanced following necessary soil preparation such as addition of organic matter, fertilizer and mulch.

Weaver (1969, p. 48-53) advised tillage for clogged Long Island basins, pointing out that this would loosen soil, overcome compaction by raindrops, and ease periodic sediment removal by inhibiting weed growth. Similarly, the Washington State Department of Ecology (1992, p. III-3-28) recommended tillage following sediment removal from unvegetated basins. Increases in infiltration rate immediately following tillage have occurred in unvegetated basins but have turned out to be temporary (Hannon, 1980, p. 46). Although tillage temporarily loosens soil, tillage and bare soil can do more harm than good in the long run. Tillage and bare soil can contribute to formation of a surface crust by suppressing vegetative cover, exposing soil to raindrops and erosion, breaking down soil aggregates, releasing individual soil particles to form a sealing layer, and preventing soil biota from building a naturally porous and conductive surface (Bianchi and Muckel, 1970, p. 27-28). Repeated tillage can form a plow pan, a polished slowly permeable layer, at the bottom of the tilled zone. Long-term reduction of infiltration rate by tillage is more pronounced on soils containing fine material than on sands.

The low hydraulic conductivity of a clogging layer can be compensated by enlarging the basin floor or adding supplemental basins upstream to eliminate inflowing runoff closer to its source.

A clogging layer can be hydraulically bypassed by puncturing it with a supplemental well or abandoning the existing basin as a settling area and constructing a replacement infiltration basin downstream.

A clogging layer can be excavated and removed to uncover uncrusted underlying soil. Weaver (1969, p. 52) advised removing accumulated sediment only when the basin floor is completely dry, after the sediment layer has mud-cracked and separated from the basin floor. In a large surface basin the layer can be placed in windrows using light equipment; maneuverability of the equipment in small areas and precise blade control can greatly reduce the quantity of material to be removed. The windrows may then be removed using a small front-end loader.

In a subsurface basin, accumulated sediment can be flushed out if the basin was constructed to anticipate this procedure. Clean water can be pumped in at one inlet, and pumped out at another. If the locations of inlets and perforated pipe and floor gradient and dimensions are favorable, sediment can be flushed out with the moving water.

Where wells are clogged by sediment, some agencies in California remove silt and fines by jetting (Jackura, 1980, p. 23). Jetting consists of partially filling a well with water, then injecting compressed air through a nozzle near the bottom of the shaft. Dirt or sand that has settled in the shaft or has clogged the casing perforations is forced out the top of the well. Jetting can also break up mineral scale blocking a casing's perforations or slots.

Microbial clogging can be treated by thoroughly drying the basin floor, allowing soil organic matter to oxidize (O'Hare et al., 1986, p. 20). Microorganisms that clog soils by working on soil organic matter are capable of producing soil-aggregating substances when subsequently dried; thus, after drying, microbiotically sealed soils can regain aggregation and high infiltration rates (Allison, 1947). Frequent dryings may be more effective than annual ones, even if the annual ones are very thorough and conducted during hot summer months. To prevent reformation of an anoxic layer, conversion of the basin's regime to a dryer one is necessary. This can be done by enlarging the floor area, increasing permeability with vegetation, or reducing runoff closer to its source in small upstream basins.

REFERENCES

Allison, L.E., 1947, Effect of Microorganisms on Permeability of Soil Under Prolonged Submergence, *Soil Science* vol. 63, pages 439-450.

Aronson, D.A., and G.E. Seaburn, 1974, *Appraisal of the Operating Efficiency of Recharge Basins on Long Island, New York, in 1969*, Water-Supply Paper 2001-D, Washington, D.C.: U.S. Geological Survey.

Behnke, Jerold J., 1969, Clogging in Surface Spreading Operations for Artificial Ground-Water Recharge, *Water Resources Research* vol. 5, no. 4, pages 870-876.

Beven, Keith, and Peter Germann, 1982, Macropores and Water Flow in Soils, *Water Resources Research* vol. 18, no. 5, pages 1311-1325.

Bianchi, W.C., and Dean C. Muckel, 1970, *Ground-Water Recharge Hydrology,* ARS 41-161, Washington, D.C.: U.S. Agricultural Research Service.

Chescheir, George M., Guy Fipps and R.W. Skaggs, 1988, Hydrology of Two Stormwater Infiltration Ponds on the North Carolina Barrier Islands, pages 313-319 of *Proceedings of the Symposium on Coastal Water Resources,* Bethesda: American Water Resources Association.

Driscoll, Eugene, Dominic DiToro, David Gaboury and Philip Shelley, 1986, *Methodology for Analysis of Detention Basins for Control of Urban Runoff Quality*, EPA 440/5-87-001, NTIS Document No. PB87-116562, Washington, D.C.: U.S. Environmental Protection Agency Office of Water Regulations and Standards.

Duley, F.L., 1939, Surface Factors Affecting the Rate of Intake of Water by Soils, *Soil Science Society of America Proceedings* vol. 4, pages 60-64.

Faulkner, Stephen P., and Curtis J. Richardson, 1989, Physical and Chemical Characteristics of Freshwater Wetland Soils, pages 41-72 of *Constructed Wetlands for Wastewater Treatment, Municipal, Industrial and Agricultural*, Donald A. Hammer, editor, Chelsea: Lewis.

Ferguson, Bruce K., and P. Rexford Gonnsen, 1993, Stream Rehabilitation in a Disturbed Industrial Watershed, pages 146-149 of *Proceedings of the 1993 Georgia Water Resources Conference*, Kathryn J. Hatcher, editor, Athens: University of Georgia Institute of Natural Resources.

Goss, D.W., S.J. Smith, B.A. Stewart and O.R. Jones, 1973, Fate of Suspended Sediment during Basin Recharge, *Water Resources Research* vol. 9, no. 3, pages 668-675.

Hannon, Joseph B., 1980, *Underground Disposal of Stormwater Runoff, Design Guidelines Manual,* FHWA-TS-80-218, Washington, D.C.: U.S. Federal Highway Administration.

Jackura, Kenneth A., 1980, *Infiltration Drainage of Highway Surace Water*, FHWA/CA/TL-80-04, Washington, D.C.: Federal Highway Administration and Sacramento: California Department of Transportation.

Jennings, G.D., A.R. Jarrett and J.R. Hoover, 1988, Evaluating the Effect of Puddling on Infiltration Using the Green and Ampt Equation, *Transactions of the American Society of Agricultural Engineers* vol. 31, no. 3, pages 761-768.

Jones, O.R., D.W. Goss and A.D. Schneider, 1974, Surface Plugging during Basin Recharge of Turbid Water, *Transactions of the American Society of Agricultural Engineers* vol. 17, no. 6, pages 1011-1014 and 1019.

Kazman, Z., I. Shainberg and M. Gal, 1983, Effect of Low Levels of Exchangeable Sodium and Applied Phosphogypsum on the Infiltration Rate of Various Soils, *Soil Science* vol. 135, no. 3, pages 184-192.

Lindsey, Greg, Les Roberts and William Page, 1992a, Maintenance in Stormwater BMPs in Four Maryland Counties: A Status Report, *Journal of Soil and Water Conservation* vol. 47, no. 5, pages 417-422.

Lindsey, Greg, Les Roberts and William Page, 1992b, Inspection and Maintenance of Infiltration Facilities, *Journal of Soil and Water Conservation* vol. 47, no. 6, pages 481-486.

Moss, A.J., and C.L. Watson, 1991, Rain-Impact Soil Crust. III. Effects of Continuous and Flawed Crusts on Infiltration, and the Ability of Plant Covers to Maintain Crustal Flaws, *Australian Journal of Soil Research* vol. 29, pages 291-209.

Muckel, Dean C., 1959, *Replenishment of Ground Water Supplies by Artificial Means*, Technical Bulletin 1193, Washington, D.C.: U.S. Department of Agriculture, Agricultural Research Service.

Nightingale, Harry I., and W.C. Bianchi, 1977, *Environmental Aspects of Water Spreading for Ground-Water Recharge*, Technical Bulletin 1568, Washington, D.C.: U.S. Department of Agriculture, Science and Education Administration.

O'Hare, Margaret, Deborah M. Fairchild, Paris A. Hajali and Larry W. Canter, 1986, *Artificial Recharge of Ground Water, Status and Potential in the United States*, Chelsea: Lewis.

Pensyl, K., and P.F. Clement, 1987, *Results of the State of Maryland Infiltration Practices Survey*, Baltimore: Sediment and Stormwater Division, Maryland Department of the Environment.

Roesner, Larry A., Ben Urbonas and Michael B. Sonnen, editors, 1989, *Design of Urban Runoff Quality Controls*, New York: American Society of Civil Engineers.

Seaburn, G.E., and D.A. Aronson, 1974, *Influence of Recharge Basins on the Hydrology of Nassau and Suffolk Counties, Long Island, New York*, Water-Supply Paper 2031, Washington, D.C.: U.S. Geological Survey.

Shainberg, I., 1992, Chemical and Mineralogical Components of Crusting, pages 33-53 of *Soil Crusting, Chemical and Physical Processes*, M.E. Sumner and B.A. Stewart, editors, Chelsea: Lewis.

Smith, Hillard B., and Luna B. Leopold, 1942, Infiltration Studies in the Pecos River Watershed, New Mexico and Texas, *Soil Science* vol. 53, pages 195-204.

Washington State Department of Ecology, 1992, *Stormwater Management Manual for the Puget Sound Basin (The Technical Manual)*, Olympia: Washington State Department of Ecology.

Weaver, Robert J., 1969, *Recharge Basins for Disposal of Highway Storm Drainage: Theory, Design Procedure, and Recommended Engineering Practices*, Research Report 69-2, Albany: New York State Department of Transportation, Engineering Research and Development Bureau.

Wilson, L.G., 1967, Sediment Removal from Flood Waters by Grass Filtration, *Transactions of the American Society of Agricultural Engineers* vol. 10, no. 1, pages 35-37.

Zobell, Claude E., 1975, Microbial Activities and Their Interdependency with Environmental Conditions in Submerged Soils, *Soil Science* vol. 119, no. 1, pages 1-2.

7

Infiltration Cases and Experiences

Infiltration is used intensely in many local areas of the United States. In other areas, experience with infiltration is non-existent. It is important for designers in any one area to learn from experiences elsewhere.

A review of some of the experiences with stormwater infiltration in different regions of the country will point out some highlights in the history of infiltration, the variety of issues it has been used to address, the variety of its responses to specific environmental and social contexts, and some causes of successes and failures. The projects described below have been documented in the literature more than cursorily or seen by the author firsthand. Some are broad public programs; others are specific site developments with or without explicit public mandates. Many other projects, about which some details are known, are mentioned in other appropriate chapters.

The first recorded case of artificial replenishment of ground water was in 1889, when the Denver Union Water Company averted a threatened water famine at its pipeline intake by spreading water over the alluvial gravel cone at the mouth of South Platte Canyon in Colorado (Muckel, 1959). In 1895, flood waters were spread over an alluvial fan at the mouth of San Antonio Canyon in California to sustain wells in the upper Santa Ana Valley (O'Hare et al., 1986, p. 3-10). These early practices acquired the term "water spreading" because their purpose was to spread water over an area of soil larger than would occur naturally, thereby increasing flow to aquifers. In the aquifers, recharged water was protected from evaporative loss while distributing itself throughout alluvial valleys where farms and cities had their water supply wells. Water spreading expanded greatly in the western U.S. during the drought years of the 1930s, when increased pumping made great demands on ground water supplies. Spreading systems can now be seen adjacent to nearly every stream in southern California, including those where infiltration and percolation rates are not ideal (Muckel, 1959).

LONG ISLAND

Nassau and Suffolk Counties on Long Island, New York have the longest and best documented experience with specifically urban stormwater infiltration in the United States (Ku and Simmons, 1986; Seaburn and Aronson, 1974). The Island's infiltration program began in 1935 as part of the Nassau County Sanitation Commission's comprehensive drainage plan. According to that plan, the use of recharge basins would conserve storm runoff by replenishing the Island's ground water, which is its only water source, and would avoid expensive public storm sewers by eliminating runoff near the source. Suffolk County adopted the concept some years later. The Island's housing boom following World War II was accompanied by a large increase in the number of infiltration basins, from 14 in 1950, to 700 in 1960, and to 2,124 in 1969. The number now is certainly well over 3,000. Studies by the U.S. Geological Survey and other organizations beginning in the 1960s have produced a well-rounded picture of how the concept is working.

Long Island's soils are derived from glacial outwash and terminal moraines and are mostly sandy and permeable, particularly on the outwash plains where a large amount of development takes place. Average annual precipitation is about 110 cm; precipitation is fairly constant during the year.

Long Island's infiltration basins were inventoried in 1969 (Seaburn and Aronson, 1974). At that time the basins ranged in area from 0.4 to 12 ha, averaging between 0.4 and 0.8 ha. Most were excavated 3 to 4.5 m below the land surface; some were as deep as 12 m. Most had overflow structures to carry excess water from one basin to another or to a nearby stream. Eighty percent of the basins drained residential areas, 17 percent drained highways, and 3 percent drained commercial and industrial areas. The average drainage area was 14 ha; the total drainage area of all the basins was 30,000 ha, or about 10 percent of the land area of Nassau and Suffolk Counties.

The floors of some basins are used for recreation by neighborhood children, either formally or informally.

The basins tend to be open, flat-bottomed excavations, usually with steep sides. Some basins include settling areas of different elevations on the basin floor; the lower level collects inflowing coarse sediment and trash, leaving the higher level relatively free to infiltrate without rapid sedimentation. Some receive the clean overflow from separate settling ponds through pipes or channels. Some include wells below the floor to penetrate impermeable strata and infiltrate directly into deeper, more permeable strata.

Sizing methods and criteria have evolved empirically (Seaburn and Aronson, 1974). A 12.5-cm rainfall is the most common design storm. Basin volume is sufficient to hold the entire runoff volume, which is found through a formula analogous to the rational formula, in which the rainfall depth is multiplied by the drainage area and by a coefficient ranging from 0.3 for residential land use to 1.0 for highly impervious industrial areas. Infiltration during inflow is not taken into account, and there is no explicit criterion for ponding time.

Runoff from streets and pavements is often delivered to the basins by local storm sewers. Although the short, small pipes are believed to be more economical than construction of long mains to collect the runoff from all development sites and convey it to outfalls at sea level (Weaver, 1969, p. 1), this amounts to building two primary local drainage systems—one conveyance, one infiltration. Open-bottomed drop inlets in the storm sewer system reduce the inflow to the basins by infiltrating some runoff close to the source. In the drainage area of one studied basin (the Deer Park basin, Seaburn and Aronson, 1974), some larger drop inlets are dry wells constructed of 2.4-m-diameter perforated concrete rings; other inlets are not larger than 1 m square and 1.5 m deep, are open only at the bottom, and clog easily.

Most basins are owned and maintained by local government agencies; some are privately owned (Aronson and Seaburn, 1974). Maintenance is variable. Where it is done, it consists mainly of collecting and removing bulk debris and cutting and removing grass on the floor of the basin.

About 91 percent of the basins are dry within 5 days after a 2.5 cm rainfall (Ku and Simmons, 1986). Those that hold water for longer periods intersect the water table, are excavated in soil of low conductivity, or are clogged with layers of sediment and debris. In some of the water-containing basins, water levels recede rapidly (within a few hours after a storm) to the pre-storm levels (Aronson and Seaburn, 1974).

In three monitored basins where the water table was 5.5 to 22 m deep (Seaburn and Aronson, 1974), small ponding events resulted in mounds in the underlying water table of about 15 cm; larger events of 5 cm rainfall resulted in peak rises of about 60 cm. The mounds commonly dissipated in 1 to 15 days, depending on the magnitude of the peak buildup and varying from basin to basin. Water table mounding has had no apparent effect on infiltration rates in the basins. However, in rare basins located near the shorelines, where the water table is shallow, ground water mounds have sometimes encroached on nearby structures.

The Centereach basin in Suffolk County (Ku and Simmons, 1986, p. 10) has been designated Ecological Recharge Basin #2 by the New York State Department of Transportation (DOT). It was constructed in 1977 to receive runoff from a 28-ha drainage area, in an area underlain by sand and gravel with the water table about 12 m deep. About 7.1 percent of the drainage area is impervious, including a four-lane asphalt highway and adjacent small stores and parking lots. The basin floor is about 0.2 ha in area; it is lined with plastic sheets so that it retains water and forms an artificial pond. DOT stocks the pond with fish and aquatic vegetation. Beneath the impermeable liner is an exfiltration system consisting of perforated pipes. When the pond level rises, the excess water flows through a filter box into the exfiltration system and ultimately into the soil and ground water.

Ku and Simmons (1986) monitored the water quality effects of five basins in five representative land use areas—strip commercial, shopping mall, major highway, medium-density residential and low-density residential. They collected runoff entering each basin and ground water below each basin, and analyzed it for arsenic, cadmium, calcium, carbon, chloride, chromium, lead, magnesium, nitrogen, phosphorus, potassium, sodium and sulfate. Concentrations of most constituents were low, generally within the standards for potable water. A major exception was road deicing salts in the winter. After movement through the vadose zone, the heavy metals were largely removed, but chloride was not. Nitrogen was sometimes in greater concentration in ground water than in stormwater due to prior contamination from septic tanks, cesspools and lawn fertilizers. No bacteria were detected in ground water beneath the basins. The authors concluded that infiltration basins are a safe method for stormwater disposal and ground water replenishment under the conditions found on Long Island, including the rapidly permeable soils. Seaburn and Aronson (1974) came to the same conclusion after monitoring for pesticides at three other basins. Pesticides and other organic material were effectively reduced or removed from the infiltrating water by the surface soil, leaving total concentrations of pesticides in the soils on the basin floors ranging from 0.4 to 40 mg/l.

Average annual recharge at each basin varies with drainage area and other determinants of quantity of runoff (Seaburn and Aronson, 1974). Evapotranspirative losses are negligible when water remains in the basins only a short time. Recharge of runoff through the 2,124 basins in operation in 1969 was estimated to be 670 million m^3/yr. This was equal to or slightly greater than recharge under natural conditions, and greater than the ground water withdrawn to supply the residents of the basin's drainage areas, based on average rates of water use on the Island (Ferguson and Debo, 1990, p. 138-139). Despite its limited water supply, Long Island can continue developing as long as its development incorporates infiltration basins that function this way.

In about 1990 all of Long Island's infiltration basins were legally designated as ground water recharge areas and are to be protected from illegal dumping by fencing. There is concern that fencing may compromise open-space values (Sarah Meyland, Citizens Campaign For The Environment, Massapequa, N.Y., personal communication, August 1990).

EARLY EXPERIENCES WITH WELLS

Roanoke, Virginia and the state of Hawaii had early experiences with infiltration wells. In both cases the decision to use wells—or any form of infiltration at all—was borne of economic necessity, if not of physical necessity. The experiences were not favorable in all respects. Their results provide ideas about adapting to specific environmental contexts, but also dangers to watch out for in doing so.

Hawaii

Hawaii's use of recharge wells demonstrates the turning to infiltration in response to changing urban design standards.

Peterson and Hargis (1973) surveyed Hawaii's early infiltration practices. In Hilo, as of 1970, 26 basins had been constructed in residential and industrial areas to compensate for poor surface drainage. Some recharge wells were constructed at low areas in streets and covered with steel gratings; the shafts were approximately 1.5 m in diameter and from 2 to 9 m deep. In the Puukapu area of the island of Hawaii, a storage basin and six recharge wells were installed in 1968. The wells were 60 cm in diameter and from 38 to 53 m deep, all terminating in unsaturated rock, probably 100 m above any water table. The County of Maui installed 17 shallow recharge wells in low-lying residential areas of Wailuku and Kahului to eliminate standing water after heavy rains, and constructed two basins 30 m in diameter and 6 m deep in Kahului.

When the County of Maui added a requirement for curbs and gutters to its subdivision code, the Kahului Development Company constructed the basin and wells shown in Figure 7.1 for its new residential area (Peterson and Hargis, 1973). Annual precipitation in the Kahului area is less than 50 cm, mostly concentrated in the winter. Intense winter storms sometimes produce considerable local runoff for short periods. The area is underlain by very permeable basalt veneered by calcareous dune sands. The basin receives runoff from a 160-ha drainage area via storm sewers. The basin is sized to hold the entire volume of the 10-year storm (250,000 m^3) in case the wells in the basin's floor fail to infiltrate at the expected rates during inflow.

The four wells are situated in pairs, in concrete chambers on each side of the outlet of the delivery pipe. The backwater location of the wells, shown by the circulation pattern in the figure, is intended to encourage deposition of sediment and debris away from the well intakes. Runoff is let into the wells through a series of screens and elbow pipes. Each well is 40 cm in diameter and 85 m deep from the finished grade, which is 1.8 m above sea level. The upper 40 m of each well is cased to separate stormwater from the fresh portion of the aquifer; storm runoff injection occurs through the lower 45 m.

Roanoke, Virginia

Roanoke, Virginia exemplifies the retrofitting of recharge wells in urban areas to solve local drainage problems, and some of the particular hazards that can arise with infiltration in karst landscapes. Roanoke is underlain by dolomite and limestone, forming a karstic topography characterized by sinkholes and few surface streams (Breeding and Dawson, 1977). The residual soil is fine-textured and deep. Precipitation averages about 95 cm/yr.

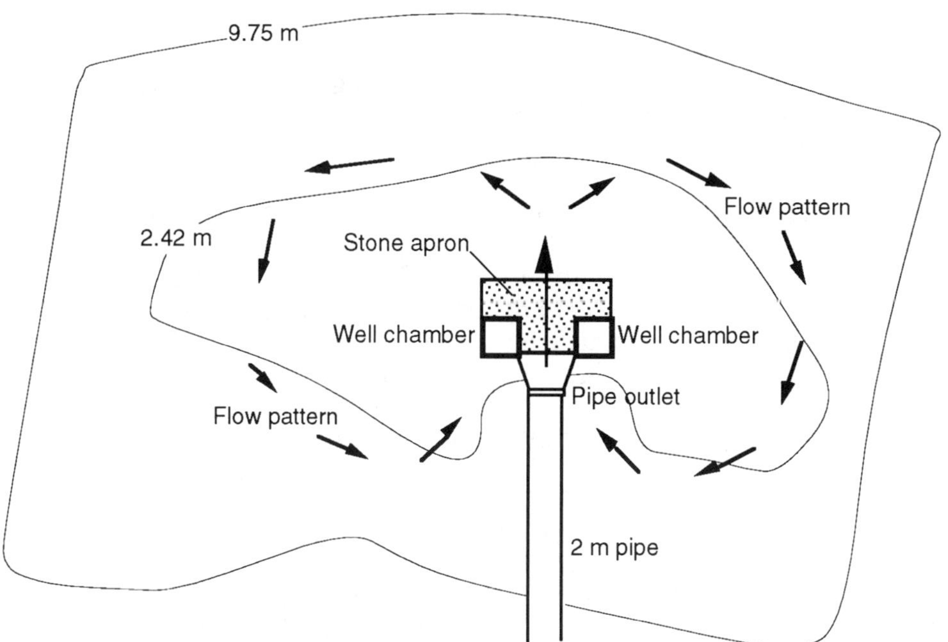

Figure 7.1 Basin with supplemental wells in Kahului, Hawaii (after Peterson and Hargis, 1973). Arrows indicate flow direction from inlet to wells. Contours show elevation above msl.

Roanoke's Williamson Road area grew rapidly in the late 1940s and early 1950s with little or no drainage planning. Urban construction filled in most of the natural ephemeral drainage ways, and ponding of stormwater in topographic depressions became a severe nuisance.

As a result, 127 wells were installed between 1955 and 1969 to drain water from streets. The construction cost was about one sixtieth that of storm sewers. The wells are 25 to 30 cm in diameter and 15.5 to 102 m deep. The infiltration capacity of individual wells ranges from 0.1 to 3.0 m^3/min.

The wells with the greatest capacities are located along fault lines which provide open conduits for water movement. Drilling encountered large cavities, indicating that some of the storm water is being discharged to ground water without soil filtration. Elevated levels of heavy metals, nitrogen, total dissolved solids and chemical oxygen demand have been found in the ground water, but have not resulted in problems to ground water users. Some wells caused ground water mounding of up to 22 m, while others were associated with drops in the water table, possibly due to opening of connecting cavities by drilling. Local topographic subsidence has been attributed to some wells.

Some wells need periodic cleaning, and some have been abandoned due to siltation. Ponding problems have remained in the vicinity of poorly operating wells.

ECONOMIC ADVANTAGES AT THOMAS MANOR

Thomas Manor in arid El Paso, Texas was one of the first developments outside Long Island incorporating infiltration for the purpose of managing urban discharge. It was designed by Glenn English, with the Federal Housing Authority overseeing objectives and criteria. The project demonstrates a direct economic advantage of infiltration in flood protection. It was cited as an example of the benefits of on-site stormwater management in the 1960s, when the concept of stormwater management was first gaining acceptance nationwide (Jones, 1967; Newville, 1967, p. 17-18).

Thomas Manor's residential lots are small and numerous. Each lot is surrounded by a garden wall. The development is located in the floodplain of the Rio Grande. It is protected from river floods by levees. The levees leave no surface outlet for drainage during wet periods, requiring some more imaginative form of drainage.

Consequently the site was designed to trap about half of the runoff during a design storm and infiltrate it on individual residential lots. Very little storm sewer was constructed. Soils are pervious; ponding following storms has been brief. Excess runoff is collected at a low point and pumped over a levee into the river. The use of the individual lots for reduction of runoff reduced the need for storm drains, conserved rainfall where it would benefit the individual homeowners, permitted a smaller sump volume, and reduced the required capacity and energy consumption of the pump.

LANDSCAPE DESIGN IN FLORIDA

From a land use and aesthetic viewpoint, Florida is home to some of the country's finest and most diverse landscape design in implementing infiltration. Infiltration is common in Florida as a protection of water quality. It is regulated at regional and local levels under the name "retention" (Livingston and McCarron, no date).

Florida's infiltration basins must deal with the state's 110 to 150 cm/yr precipitation. They are usually surface excavations, dispersed among small urban site and watersheds and planted with turf or native lowland vegetation. They are sized for a given volume of watershed runoff, such 2.5 cm, in order to capture the highly polluted first flush of runoff from each storm. Design practice has tended not to include checking the basins for storm ponding time or long-term standing water, and many of them have become ephemeral ponds or wetlands. In this wetland-rich state urban residents are used to seeing seasonal wetlands as part of the natural environment. The effect of any one basin partly failing to capture the required volume is masked by the concept that stormwater passes through a "train" of many management practices.

The 92-ha Maitland Center office park just north of Orlando exemplifies both a smooth integration of infiltration basins with urban design and a casual degree of hydrologic success. It was constructed in the 1980s following plans by the multidisciplinary firm of Post, Buckley, Schuh and Jernigan. Impervious cover is over 50 percent. The edges of the property were set aside for shopping, residences and hotels. The remainder of the property was divided into office lots served by roads and utilities; individual developers laid out each lot for

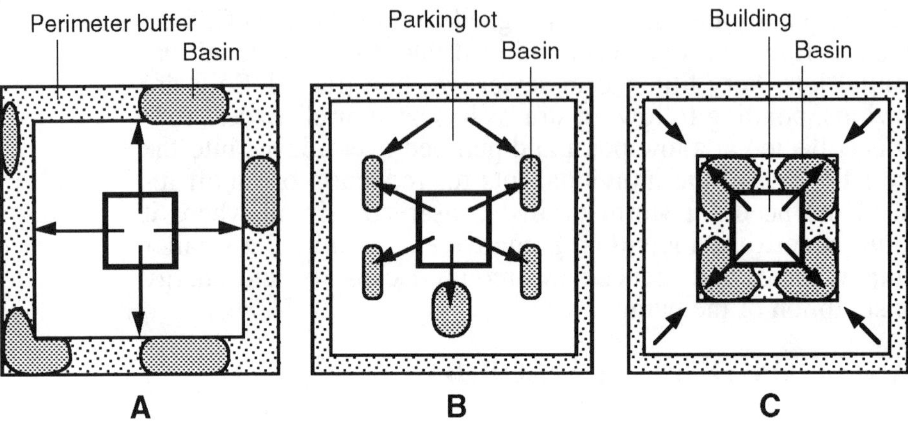

Figure 7.2 Types of locations of infiltration basins within development lots in the Maitland Center office park near Orlando, Florida. A, in perimenter landscape buffer strips; B, in grass islands in parking areas; C, alongside buildings. Arrows indicate typical flow directions.

buildings and parking. Protective covenants established criteria for landscaping, drainage, etc. A 10 m wide landscape buffer lines the perimeter of each lot. Sidewalks and systems of berms and plantings for screening parking areas wind through the buffers and road rights of way.

In each office lot, infiltration basins have occurred in the three types of locations shown in Figure 7.2: in perimeter landscape buffer strips, in grass islands in the middle of parking lots, and alongside buildings, between the buildings and the parking areas. The locations chosen and detailed forms within those locations varied with property and building configurations, the needs of circulation, etc. The primary drainage system confines the design flow within basins. During larger storms, excess water overflows basins through pipes or grassed swales, connecting the basins in a series downgradient to sinkhole lakes at the lowest elevations of the site.

Many of Maitland Center's basins were laid out with turf in mind (usually St. Augustine grass). On about half of the basin floors the turf has been killed by saturated soil or standing water that half-fills the basins each season. The basin shown in Figure 7.3 parallels the side of an office building adjacent to sidewalks and the building's main entrance. Wetland vegetation such as cattail that volunteered here has been suppressed by property managers because of fear of snakes, leaving maintained turf essentially to the edge of the standing water. Nevertheless, ducks and egrets have been seen in the turf-edged basin.

A more complex and structural drainage system is in the Sunbay Club, one of Maitland Center's residential areas. The 14-ha area contains 359 residences in low- and mid-rise condominium clusters. Runoff is let into a series of small ponds with sealed bottoms, often directly abutting buildings and containing fountains or other features. In the ponds solids settle out and the water is aerated. Primary drainage ends in a gallery of perforated pipes under tennis courts, as shown in Figure 2.14.

Figure 7.3 Infiltration basin in Maitland Center, near Orlando, Florida. An office building is on the left; sidewalks in the street right of way are in the foreground and on the right.

MARYLAND'S STATE-WIDE PROGRAM

In the 1980s, Maryland had the country's most creative initiative in infiltration and in stormwater management in general. This can be attributed at least partly to the energetic leadership of the state's Stormwater Management Division by a man named Earl Shaver. The state's infiltration program was aimed at an ambitious range of environmental objectives including storm flow volume reduction, base flow augmentation, water quality improvement, ground water replenishment and combatting of saltwater intrusion into aquifers.

In 1984 the state adopted a set of standards and specifications (Maryland Water Resources Administration, 1984) that required developers to consider infiltration on sites with soil at least as conductive as silt loam (0.68 cm/h). The document was founded admirably in the state's physiography and natural processes, but showed surprisingly little awareness of the extensive prior experiences in California and on Long Island. According to the document, basin volumes were to be sized to capture the increase in storm flow volume attributable to development. The frequency of the storm and the allowable duration of ponding varied with land use context and type of basin.

By 1992, about 1,000 infiltration basins of various types had been built in Maryland (Lindsey, Roberts and Page, 1992).

One of these was at the Cromwell Field shopping center in Glen Burnie (Charles Brenton, personal communication, 1987). On the surface the site looks like a very conventional strip shopping center with well over 90 percent impervious cover. The only visible sign of a drainage system is conventional-looking drop inlets in the parking lot. However, the inlet walls are perforated to feed into a stone-filled basin under the parking pavement 2 m deep, 0.4 ha in area, and surrounded by a filter fabric. Despite the extensive impervious cover—normal for many commercial sites—returning water to the underlying soil and ground water has maintained the hydrologic function of the site.

Adjacent to the Cromwell Field site is the office of Patio Enclosures, Inc. The small impervious parking lot drains through curb cuts into a small grass bowl adjacent to the highway, where runoff is contained and infiltrated. A small concrete flume carries runoff down the side to the floor. There is no dead grass or other sign of overly saturated soil. This is the end of the small site's drainage system—there are no other inlets, pipes or other drainage structures, except for overflow into the highway system during very large storms. The little basin illustrates the potential simplicity and economy of planned on-site drainage.

In about 1991 a coalition of agencies including the Maryland Department of the Environment constructed a demonstration stormwater management area at Fairland Regional Park near Laurel (Berg and Clement, 1992). It includes several subsurface infiltration basins. All the stormwater features are explained by informational signs. Although the collection of control structures is more like a single-purpose museum than a functionally integrated site development for human use, it is a valuable educational resource for designers working in or visiting the region.

Three years after Maryland's infiltration program started, Pensyl and Clement (1987) observed basins constructed under the program. A third of the surveyed basins were found to hold some standing water, preempting the basins' capacity to capture further flows and, in some surface basins, creating a nuisance and a safety hazard. These results were essentially confirmed by a followup study five years later (Lindsey, Roberts and Page, 1992). The researchers attributed the ponding to sedimentation, inadequate maintenance, inadequate soil conductivity data in design, and compaction during construction. None of the researchers reviewed the adequacy of as-built basin dimensions for ponding time or long-term standing water, nor did they consider in-place soil crusting as a possible cause of ponding.

These results dampened the enthusiasm of the state's staff for infiltration. In addition the state's initiative in stormwater management in general declined with an administrative reorganization of the relevant agencies and the moving away of Earl Shaver to another state. Nevertheless, the 1984 standards and specifications are still in effect. With further study and refinement of the program, perhaps the number of failures could be reduced and the potential benefits of infiltration could be fully realized.

GROUND WATER EFFECTS IN ARID REGIONS

The following two projects attempted to use infiltration to extend the useful life of fossil ground water in arid regions. Although such atempts can be admirably clever and aggressive, their potential success is limited ultimately by the self-renewing supply of water in precipitation.

Dell City, Texas

Dell City has extended ground water supplies to a degree by using infiltration to help tame the violent, destructive forces of flash floods.

In the arid Dell Valley of western Texas, rainstorms are infrequent but occasional thunderstorms are intense (Jensen, 1990; Logan, 1990). Floods originating in nearby hills flash across alluvial fans and plains without defined channels, creating annual flood losses and one disastrous flood in 1966.

At the same time ground water levels have been falling because of heavy irrigation usage in agricultural lands. Land in the valley was used for grazing until 1947, when the discovery of abundant fossil ground water at depths of 60 to 450 m brought about a rapid change to irrigated farming. As of 1990, the valley's 16,000 ha represented the biggest concentration of irrigated cropland in Texas.

The water pumped from the aquifer is moderately saline, suitable for agriculture but not for municipal use in the town of Dell City without desalinization. The water quality has been deteriorating with time, probably because of irrigation water's leaching of salts and subsequent percolation back to the water table. In the center of the valley is a topographic basin, the Dry Salt Lake, where naturally occurring salts have concentrated by evaporation. The cones of depression around irrigation wells risk drawing highly saline water from the core of the basin into the inhabited edges of the valley.

The U.S. Soil Conservation Service proposed building four flood control dams at the feet of the mountains, one at each of the major intermittent streams that discharge into the valley. However, because there are no defined rivers in the area, it was difficult to drain the reservoirs following a flood without again causing flood damage. Conveyance of surface water all the way to the salt basin at the bottom of the valley would be prohibitively costly.

The solution was to install a series of recharge wells downstream from each dam. According to the current design scheme, runoff will flow into the reservoirs during storms and then flow to the wells at controlled rates, transforming captured flood waters into ground water supplies. The system will add about 24 million m^3 of fresh water to regional aquifers each year, helping to stabilize ground water levels, reduce energy costs of pumping and decrease salinity.

As of 1990 two dams were completed; 11 gravity-operated wells 50 cm in diameter, from 365 to 450 m deep, were installed and tested, and design of the conveyances from dams to wells was underway. Well locations were selected using photogeologic techniques; well recharge capacities were developed with acid treatments.

High Plains, Texas and New Mexico

The High Plains experience demonstrates the limits of infiltration, or of any other type of water management, in solving ground water problems where ground water consumption simply exceeds total possible supply.

The Ogallala is the principal aquifer below the semiarid High Plains of western Texas and adjacent New Mexico. The region's farms were severely eroded during the Dust Bowl of the 1930s. Today they produce corn, beef, cotton, wheat, sorghum, alfalfa and soybeans that account for a substantial portion of the country's foreign exchange. The price of today's prosperity is that some 50,000 wells are now withdrawing Ogallala water for center-pivot irrigators faster than recharge replenishes it (Zwingle, 1993).

The High Plains surface in this area is peppered with more than 17,000 ephemeral playa lakes in shallow depressions formed by eolian polishing of an alluvial plain (Jensen, 1990). On average there are one or two playas per 250 ha. Lakes vary in area up to 80 ha and in depth from 1 to 3 m when full. The lakes receive rainfall and runoff during storms, particularly thunderstorms in the spring and fall. Almost all playas have clay bottoms. When wet, the clay swells, restricting infiltration. Fifty to 80 percent of the ponded water is lost to evaporation, leaving only small recharge. Relatively rapid infiltration and recharge occur in the rare episodes when water levels rise to the sandy areas at the margins.

In an attempt to reduce losses and increase recharge, researchers at Texas Tech University and elsewhere developed prototype projects for draining water from playa basins into the aquifer. The prototypes passed water through filter beds to remove suspended solids and microorganisms, and then into gravity wells. Results showed that wells that punctured an aquiclude to recharge the main aquifer could reclaim 60 to 80 percent of the water formerly lost to evaporation (Jensen, 1990).

However, the magnitude of the ground water decline exceeded infiltration's ability to reverse it economically or even physically. In recent years, average water table decline has been slowed, but only by closing down of some irrigators for which deep pumping costs had become uneconomical, return of their farms to dryland farming, conservation of water and energy in the remaining irrigated fields, and regulations on spacing of new wells.

The opposite type of ground water problem—a rising water table in the shallow aquifer—occurred in the city of Lubbock, where some playa lakes were altered to make room for urban development (Jensen, 1990). The playa depressions were deepened and the clay surface layer excavated, reducing the ponding areas and increasing infiltration rates. The increased recharge caused ground water levels to rise an average of 0.6 m every year since 1965, faster in some locales. A number of playas became perennially full of water and began to overflow when rainfall occurred. Foundations and basements of buildings on the Texas Tech University campus and elsewhere have been threatened. Many of the alternatives suggested to deal with the excess water have been based on increased pumping to stabilize the ground water level. The water could be used for irrigating parks and medians, discharged to lakes for recreational use, or treated and used in the city water supply. These proposals have yet to be implemented.

INFILTRATION IN COLD CLIMATES

Two projects in Canada illustrate infiltration's application in cold climates.

Calgary, Alberta

Buekeboom (1982) applied infiltration in Calgary, Alberta. The site was a new dinner theater requiring largely impervious surfaces, but having no storm sewer or other public discharge

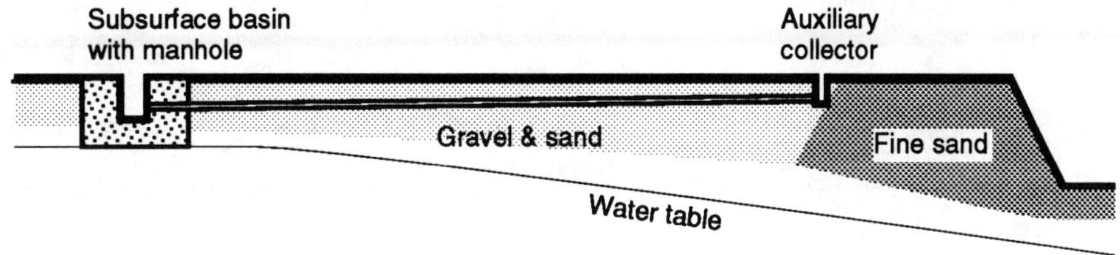

Figure 7.4 Site of a dinner theater in Calgary, Alberta (after Buekeboom, 1982).

point. Figure 7.4 shows the site, located on a topographic terrace underlain by fine sand and bounded by a 7-m-high escarpment. In order not to saturate and destabilize the slope, infiltrating water had to be kept away from the escarpment. A basin was constructed under the parking pavement; an auxiliary inlet fed surface runoff from the escarpment area to it by a pipe. To take snowmelt into account, the design storm volume was doubled, giving 138 m^3 required capacity. Borings gave hydraulic conductivities and ground water levels. Dispersion of water from the basin was calculated to prevent water level rise from intercepting the base of the escarpment. No hydrologic or pavement problems resulting from freezing in the basin have been reported.

Ottawa, Ontario

In Ottawa, Ontario, in 1985, the city government relocated Riverside Drive into a 20-year-old landfill site (Johnson Sustronk Weinstein and Associates, 1983). The four-lane divided highway gives access to Rideau River Park. Storm sewers were not permitted because of potential methane buildup from the landfill material. Part of the landfill material was excavated, relocated off the roadway, and replaced with surplus rock fill from other projects. Figure 7.5 shows the roadway construction. The void space in the rock provided a reservoir for roadway drainage. The median was surfaced with open river stone for drainage into the rock subbase. Supplemental drop inlets connected to perforated pipes in the rock in the event inclement weather would seal the river stone. After two years of operation

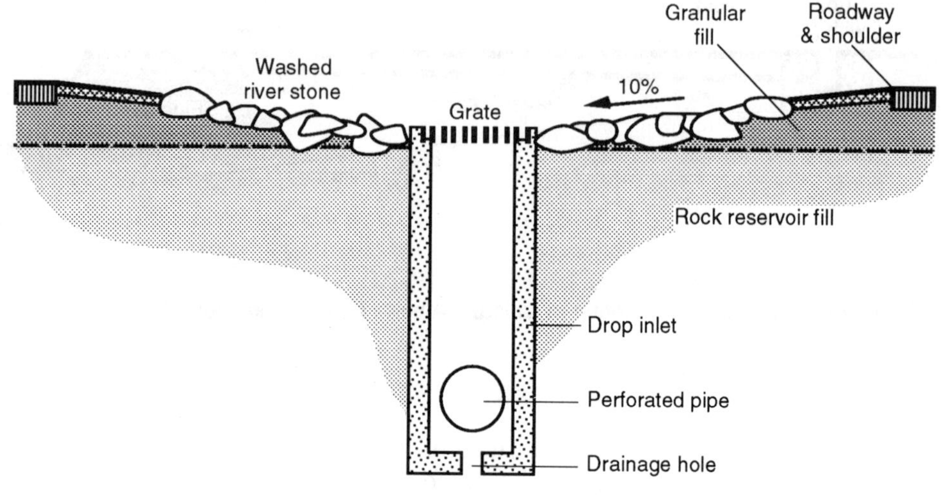

Figure 7.5 Construction of Riverside Drive, Ottawa, Ontario.

the project had experienced no problems with drainage or stability. However, the designers anticipate that flushing and cleaning of the river stone and inlets may be required in the future to maintain rapid drainage (W.S. Beveridge, Regional Municipality of Ottawa-Carleton Transportation Department, personal communication, February 18, 1987).

PLANNING STUDIES

Although planning studies do not provide actual experience with constructed basins, examples of them can be instructive in how to analyze sites, develop standards, communicate objectives, and connect runoff management to future land use and general environmental management.

The leader in environmental planning since World War II has been Ian McHarg of Philadelphia. His planning studies have been landmarks in planning analysis and methodology. Among the range of environmental objectives that those studies promoted was stormwater infiltration (referred to in the studies as

Table 7.1 Basin areas required to infiltrate excess runoff due to development in Medford Township, New Jersey (Juneja, 1974, p. 29). HSG is SCS hydrologic soil group. The symbol "—" means that a combination of HSG and infiltration rate did not occur in the township and so was not calculated.

		Prospective 0.2 ha residential					Prospective 0.1 ha residential				
			Site area required for each infiltration rate					Site area required for each infiltration rate			
Existing land use	HSG	Excess runoff (cm)	>16.0 cm/h	5.08 cm/h	1.60 cm/h	0.508 cm/h	Excess runoff (cm)	>16.0 cm/h	5.08 cm/h	1.60 cm/h	0.508 cm/h
0.4 ha	A	0.05	0%	0%	—	—	0.30	0%	2%	—	—
resi-	B	0.10	0%	0%	2%	—	0.61	1%	4%	13%	—
dential	C	0.10	—	1%	2%	7%	0.48	—	3%	10%	32%
	D	0.13	—	—	3%	—	0.58	—	—	12%	—
0.2 ha	A	0.00					0.25	1%	2%	—	—
resi-	B	0.00	No increase in runoff;				0.51	1%	3%	11%	—
dential	C	0.00	no infiltration required				0.38	—	3%	8%	25%
	D	0.00					0.71	—	—	15%	—

"retention" or "recharge") as a way to balance, maintain and restore natural hydrologic processes. Two of his most widely cited studies were Medford and The Woodlands.

Medford, New Jersey

The study for Medford Township, New Jersey was probably the first planning study adopting a public policy of infiltration for the strictly environmental purpose of maintaining landscape flow regimes. The study was developed by McHarg's group at the University of Pennsylvania (Juneja, 1974). Medford is in the Coastal Plain, with low relief, permeable sandy and medium-textured soils, and perennial streams. In the early 1970s the township was almost half wooded in pine barrens vegetation, but developing rapidly.

In the township's planning report, infiltration was specified for all development that increased runoff. The excess runoff due to development during the 10-year storm was to be infiltrated within three hours of ponding time.

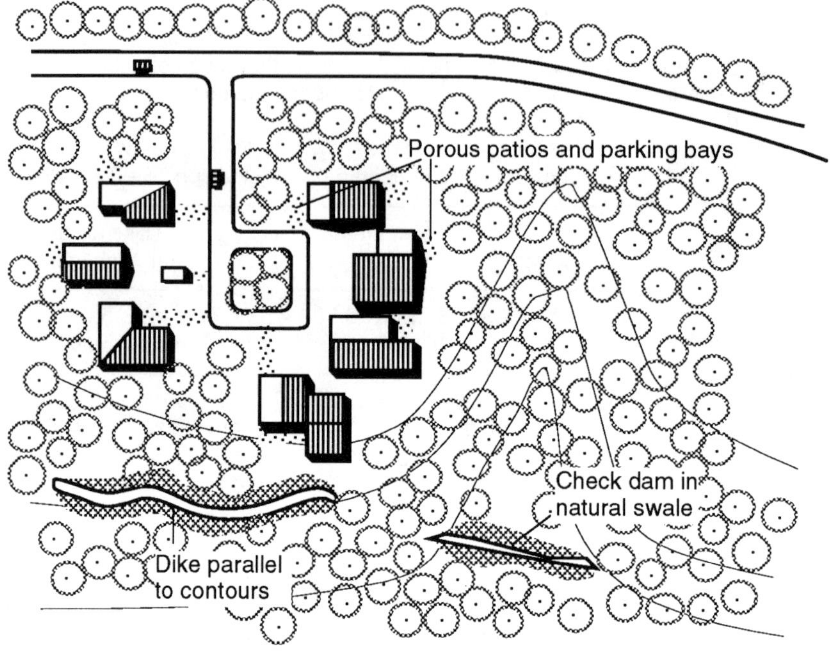

Figure 7.6 Cardinal Ridge in Medford, New Jersey, designed by Interface Environmental Conultants (after Palmer, 1981, p. 314).

In order to plan for infiltration in specific future land uses, depth of direct runoff during the storm was estimated by applying assumptions of impervious cover to the SCS method. Increase in runoff was calculated for every combination of existing land use, prospective land use, and SCS hydrologic soil group. The basin floor area required to infiltrate the increase within three hours was calculated as a fraction of lot area for each of four categories of infiltration rate cited in the local soil survey. The results were presented in a table, a portion of which is reproduced in Table 7.1.

One of the first sites to be developed following the plan's adoption was Cardinal Ridge (Palmer, 1981, p. 97-119 and 306-333). The 44.5-ha site was unusually hilly for Medford, wooded with pine and oak, with three ponds along a headwater stream. Soils were porous; the water table was more than 1.5 m deep over most of the site. The site plan by the interdisciplinary firm of Interface Environmental Consultants consisted of 77 detached homes on 0.2-ha lots, arranged in clusters around small access roads to minimize impervious surfaces. Half of the site was left in woods and ponds. Roads followed the contours to prevent direct runoff into ponds. Check dams and infiltration areas were placed in roadside swales to prevent runoff from roads, driveways and parking areas from being carried into lakes. A 30-m zone of undisturbed vegetation protected the edges of lakes and streams from compaction, erosion and pollution.

Figure 7.6 illustrates Cardinal Ridge's clustered dwelling units, small impermeable paved surfaces, and natural vegetation maintained in yards and open space. Houses were situated to minimize removal of existing tree roots and forest canopy. Driveways and parking areas were paved with permeable material such as gravel. Outdoor patios were of permeable materials such as wooden decks. Deed restrictions controlled tree removal, yard plantings, and use of fertilizers. Dry wells controlled volume of runoff; swales controlled its direction. Below housing clusters, infiltration areas were formed with check dams in natural swales and with dikes paralleling slope contours.

The Woodlands, Texas

McHarg's firm of Wallace, McHarg, Roberts and Todd developed the ecological basis of a master plan for the 7,300-ha The Woodlands, a new town near Houston, Texas (Sutton, 1974). The plan's ecological underpinning functioned simultaneously as an environmental impact analysis. Submission of the plan by Mitchell Energy and Development Corporation to the U.S. Department of Housing and Urban Development in 1972 was the basis for a 50 million dollar loan guarantee.

The Woodlands' Coastal Plain site was flat and chiefly in pine-oak forest (McHarg and Sutton, 1975). Annual precipitation averaged about 114 cm. Much of the soil was poorly drained. One third of the site was within the 100-year floodplain of three creeks. The streams flowed in broad, shallow floodplains and were characterized by low base flows and high peak flows.

Land uses were distributed according to a combination of permissible coverage and permissible clearing (McHarg and Sutton, 1975). Permissible coverage was the fraction of an area

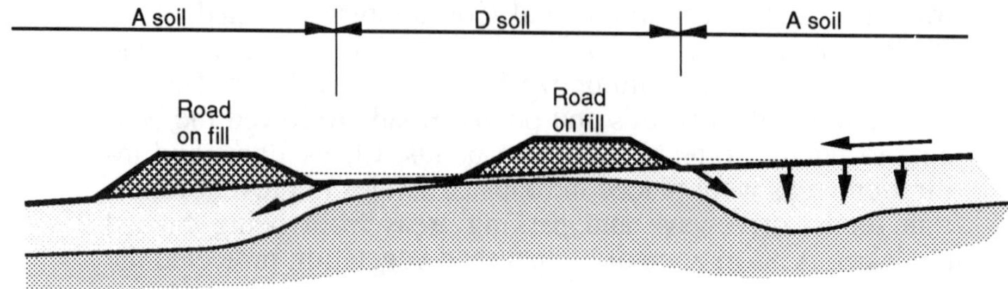

Figure 7.7 Use of road berms to induce infiltration into permeable soils at The Woodlands, Texas (after Sutton, 1974).

that could be rendered impervious without affecting the ability of the remaining soil to absorb precipitation and runoff from a small (2.5 cm in 6 hours) storm. Permissible coverage was based on the excess storage capacity of each soil, determined by the depth of the soil above the seasonal high water table. SCS hydrologic soil group A had 90 percent permissible coverage, B 75 percent, C 50 percent, and D 100 percent (effectively impermeable under present conditions). Permissible clearing was the fraction of an area from which vegetation could be removed; for each vegetation type this reflected its quality and landscape value, its importance as a wildlife food source, and its tolerance to compaction and development activities. Permissible coverage and permissible clearing were combined to make an index of landscape tolerance.

The amounts of impervious surface and clearing required for each prospective land use were estimated. Land uses having extensive coverage and clearing impacts were matched with soils of low permeability and vegetation of low quality; those having slight impacts were matched with soils of excess storage capacity and vegetation of high quality.

The first phase of development, called the Village of Grogan's Mill, was 790 ha in area (McHarg and Sutton, 1975). The 25-year floodplains of large streams and 100-m drainage easements were set aside as open space, leaving 510 ha developable. Land uses were distributed in the developable area based on landscape tolerance, configurations of land use units, proximity to amenities, marketing potential and construction sequence. The resulting development included 4,350 dwelling units, community facilities, a business park and a commercial leisure center. The planners produced a manual for site designers for complying with the plan and preserving the natural drainage system and woodland environment. The manual matched each site condition with an appropriate housing configuration, foundation type, setback dimensions, access, parking, and yard space.

Drainage was planned to utilize and enhance the site's natural vegetation, soil litter layer, soil storage capability, swales, streams and ponds to retard erosion, promote infiltration and recharge, increase base flow, protect water quality, enhance scenic amenity, maintain the water table, diminish runoff, and protect natural vegetation and wildlife habitats. Figure 7.7 shows a divided road with a median, aligned along the contours and raised on fill to form check dams enhancing infiltration into permeable A soils. Small check dams were to be placed in series where swales crossed A soils. Detention reservoirs were installed in the main streams. Major roads and the most intense development were sited on ridges away from drainage areas. An experimental parking lot with porous paving was constructed at Grogan's Mill's commercial leisure center. The Woodlands' open drainage system saved more than 14 million dollars in construction cost compared with storm sewers.

The Woodlands plan has been updated as development has proceeded. Over the years a variety of planners, landscape architects, engineers and hydrologists have been involved both with the master plan and with individual construction phases.

CALIFORNIA

Since developing the country's first artificial ground water replenishment a century ago, California's public works agencies and water rights holders have developed an extensive and elaborate system of ground water manipulation and utilization. The system has had favorable water conservation effects when each feature in the system is considered alone. However, most features have been single-purpose, optimizing water consumption for one user while disregarding the potential for broader environmental or social benefits. The features have tended to be large, central public works, managing concentrated volumes of water at focal points while the general landscape is impoverished. California is currently attempting to escape from the bounds of central management for special interests, and emerge to a broader concept of the environment in which all the landscape and all of the landscape's inhabitants participate.

The city of Modesto exemplifies the extensiveness of runoff management in a medium-sized California city. As of 1980, the public streets in 70 percent of the city's area were drained by 6,500 rock wells directly into the shallow aquifer (Hannon, 1980, p. 16-17). The wells were like that in Figure 7.8. They were typically 76 cm in diameter, 6 to 12 m deep, lined with corrugated metal pipe and filled with stone. Certain wells have overflowed onto city streets despite recurrent maintenance; the cause has not been reported.

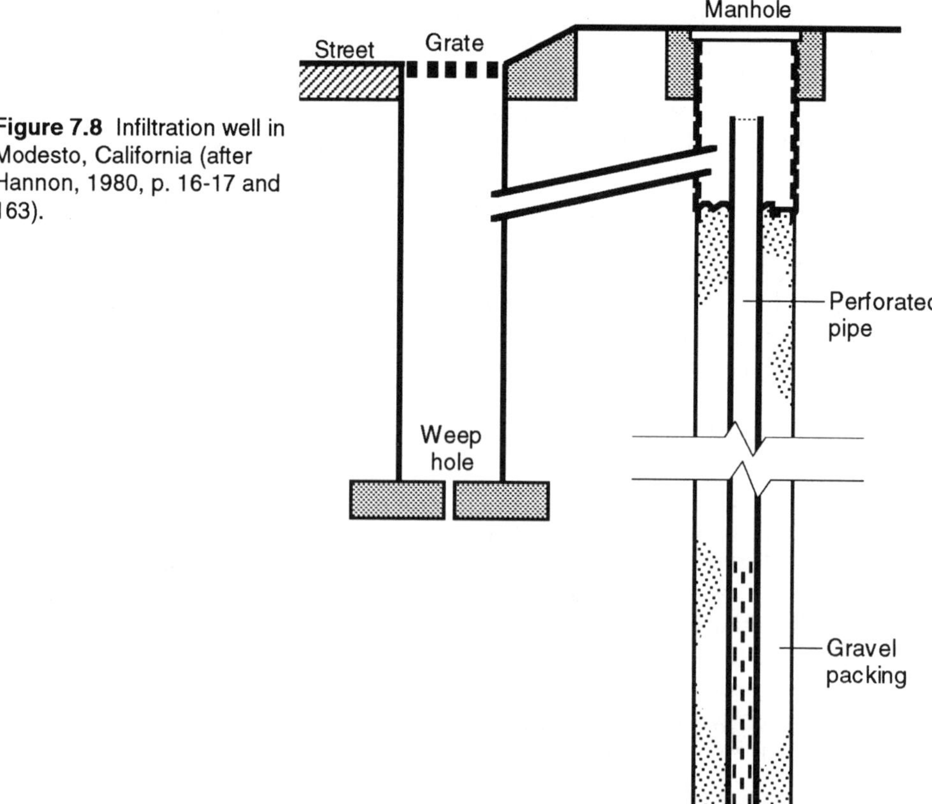

Figure 7.8 Infiltration well in Modesto, California (after Hannon, 1980, p. 16-17 and 163).

Los Angeles

Los Angeles has California's most intense system of urban ground water manipulation. Annual rainfall in the city is low (31 cm) and concentrated in the winter season. Stream flow is seasonal and intermittent. Ground water in the alluvial sediments under the city could not possibly support the metropolitan region's 14 million people without artificial supplement and management.

Rainfall in the San Gabriel Mountains above the city can be as much as three times greater than that on the plain where the city is built. All the streams emerging from mountain canyons are seized upon with dams for flood control and water supply. At the dams, diversion channels extract some water for immediate use. Just below the mouths of the canyons, spreading basins recharge the dams' gradual discharge into the ground water in gravel alluvial fans to supply nearby wells.

Farther down the stream system, the streams cross the outcrop areas of the major coastal plain aquifers that act as the city's immediate water supply. Here the Los Angeles County Flood Control District has developed vast spreading grounds since about 1930, often with the cooperation of the Army Corps of Engineers. The initial purpose of the grounds was to augment the recharge capacity of natural stream beds. Today the District diverts stream water, urban runoff and reclaimed effluent from stream channels and drainage canals and recharges them together. In the aquifers water is stored for later municipal, industrial and agricultural use. The spreading grounds total 1,300 ha, annually infiltrating 9 million m^3 of water, valued at more than 20 million dollars (O'Hare et al., 1986, p. 76-77). Individual sites range in area from 2 to 230 ha; the largest is the Rio Hondo spreading ground. Water is fed to the grounds by 19 dams and over 3,200 km of drainage channels. The outflow gates of flood control dams are operated on a schedule to permit nearly full recapture of floodwaters in downstream basins. Although runoff in Los Angeles is seasonal, the spreading grounds operate year-round to infiltrate imported Colorado River water and reclaimed wastewater. Flocculants and settling are used to reduce suspended solids entering the infiltration areas. Alternate wetting and drying has been found necessary to maintain infiltration rates at most sites; this has required the provision of additional, standby basins so spreading could be rotated from one to another. Periodically silt deposits have been removed to restore infiltration rates.

Fresno

The city of Fresno, in California's alluvial Central Valley, has developed a public drainage system with elaborateness typical of California, but with more attention to multiple use and the natural environment. Fresno's terrain is flat and has few natural channels. The average annual rainfall of only 28 cm is concentrated into the winter months. The Fresno Metropolitan Flood Control District utilizes infiltration basins to control urban flooding and reduce requirements for downstream channels and storm sewers (Jackura, 1980, p. 1; Nightingale, 1978 and 1988). As of 1988 the district had 90 basins, annually receiving about 11 million m^3 of urban runoff. Most basins are planted in Bermuda grass. Some are developed for recreation such as baseball and soccer. Some basins have subbasins excavated to depths up to 8 m, and some have gravel-filled recharge shafts. The unturfed basins annually receive an additional 18 million m^3 of imported water for year-round recharge. Five basins studied by Nightingale (1987) had areas of 2.4 to 4.0 ha, depths of 2.4 to 8.8 m, storage capacities of 50,000 to 250,000 m^3, and depths to water table of 11 to 26 m. Their catchment areas ranged in size from 163 to 937 ha and were more than 80 percent urban residential.

The California Department of Transportation has installed over 100 drainage wells in the Fresno area since 1935 (Jackura, 1980, p. 16-19). They were installed to remove standing water from isolated areas where no other alternative was viable. The sandy substrate west of town has provided good drainage for basins and wells located along Highway 99. The water table is more than 30 m deep; above the water table are sandy, dry strata separated by silt or hardpan layers. The average well is about 10 m deep and clears the water table by 3 m or more. Many different designs have been used, and several mechanical and chemical maintenance methods have been applied with varying results. Satisfactory infiltration has been experienced by installing wells and trenches in basin floors to penetrate impervious strata. Sediment and residues of waste oil dumped from nearby gas stations have reduced the performance of some wells.

At Fresno's largest facility, called Leaky Acres, seasonally excess surface water from mountain runoff is infiltrated to augment the local ground water supply (Pettyjohn, 1981). Ten basins occupy 47 ha on an alluvial fan composed of heterogeneous deposits and containing measurably saline ground water. Water originating in the Kings River has been infiltrated about 300 days per year since the early 1970s. Recharge with this fresh water has noticeably reduced the salinity of ground water for a distance of about 2 km in the direction of regional ground water movement. Turbidity of well water in the vicinity of the project temporarily increased after the project was started; the increase was attributed to extremely fine colloids that were leached from the surface soils. The vegetation, bird life and insect nuisances of Leaky Acres have been studied and are described in Chapter 2.

Village Homes

A new kind of approach was taken at Village Homes, a 28-ha community of 224 homes in the college town of Davis, in California's Central Valley. Annual rainfall in this area is low (43 cm) and concentrated in the winter season. In the first part of this century, water rights holders and state and federal agencies had brought water in aqueducts from distant mountains for irrigation and municipal use. Urban stormwater was conventionally discharged through underground pipes into rivers (Thayer and Westbrook, 1989).

The Village Homes site was planned and developed in the 1970s by the self-taught designer-developer, Michael Corbett. Corbett gave the community parallel systems of streets and open space corridors. The site's natural stream channels had been obliterated when the land was leveled to facilitate gravity-driven agricultural irrigation. Consequently, for the site's new residential use, new drainage channels were created artificially by grading to conform to the open space plan (Corbett, 1981, p. 88-89). Corbett worked closely with his engineer on the grading and drainage plan and spent much time on the site during construction to direct finish grading (Thayer and Westbrook, 1989).

Lots were graded so that roofs and lawns drain to shallow swales running through the open space corridors behind the houses (Corbett, 1981, p. 88). Although front yards drain to the narrow street pavements, the streets in turn discharge into surface swales through notched curbs. Compared to closed structural systems, open drainage with water in contact with the soil was believed to increase vegetative variety, reduce need for irrigation water for appropriately selected plants along the drainage way, reduce velocity, decrease downstream peak flow and runoff volume, increase infiltration, support wildlife habitat, provide human amenity, symbolize interaction with nature, and require little single-purpose maintenance (Thayer and Westbrook, 1989).

The open swale system drains across the site's gentle topography with little drop in elevation. Swales are landscaped like seasonal stream beds, with rocks, shrubs and trees. Small wooden bridges span the swales for foot and bicycle traffic. The residents of one cluster of eight houses decided to maintain their part of the drainage system as a shallow pond year-round by sealing the bottom, adding city water during the summer, and allowing cattails and other marsh vegetation to develop (Thayer and Westbrook, 1989). The pond is stocked with mosquito fish to suppress insect larvae. Other homeowners have planted riparian trees and herbaceous plants along the drainage way. Frogs, toads, newts, blackbirds, flickers, owls, doves, killdeer, opossum, skunk, raccoons and numerous butterflies and moths have been identified in the drainage way system.

Small infiltration basins, called percolation ponds, were formed by widening the swales in areas of well-drained gravelly soil and installing wooden or rock check dams 15 to 20 cm high (Thayer and Westbrook, 1989). The floors of various percolation ponds are surfaced with lawn, clover, coarse sand, gravel and river cobbles. Undesired weed growth in gravel surfaces has required maintenance. A typical percolation pond is shown in Figure 2.10. Excess runoff spills over the small weirs and flows down the swale system into further series of basins. Ponding depth never exceeds 20 cm; thus ponding time after a storm has seldom exceeded four hours. The system was designed to absorb a 10-year storm on the site, without discharge to the preexisting city storm sewer. As of 1989 the system had not discharged any overflow water into the city storm sewer during the community's 13-year life (Thayer and Westbrook, 1989).

Overcoming institutional conventions to build the open drainage system required resolute persuasion of local commissions (Thayer, 1977), but it saved $800 per house compared to structural storm sewers (Corbett, 1981, p. 87). In the early 1980s mandatory tax cuts reduced the funds available for Davis' storm sewers and other public works. The city subsequently adopted a policy that followed Village Homes' lead by requiring that future new development was to have no downstream impact, either in total volume or peak flow rate, during given design storms (Thayer and Westbrook, 1989).

Etiwanda-Day

John Lyle's landscape group at California State Polytechnic Institute in Pomona has suggested in numerous planning studies that all of California's water management could become more environmental. Their plan for future land use in the Etiwanda-Day area in southern California (Carlson et al., 1986) is an example of the kinds of benefits that await more conscientious landscape design.

In this 1,440-ha area the Etiwanda and Day Canyons emerge from the San Gabriel Mountains. The major landowner at the time of the study was the University of California, which was interested in realizing the full revenue potential of its land through urban development, while setting aside Day Canyon for biological research. In planning for increased urban density, the planners sought to protect the area's environmental and aesthetic qualities sustainably, and thus to unite the man-made and natural environments.

Etiwanda-Day's mountain slopes are covered with chaparral incised by the unique riparian communities of the two canyons. The canyons' streams drain onto alluvial fans and plains, where the coarse, highly conductive soil is covered with coastal sage and scrub. Parts of the site are used by public agencies for power transmission, water treatment, and maintenance of National Forest land. Implementation of any proposed development would require cooperation among the various owners and interest groups.

On the alluvial fans, water agencies use spreading basins to recharge runoff and imported water into ground water for later withdrawal and use. In southern California imported water has been politically allocated and increasingly expensive. The region's urban runoff drains into Los Angeles' vast public works system as described above. The region's structural, centralized storm sewers, channels and flood control basins tend to be detrimental to headwater peak flows and stream base flows, water quality, wildlife, and the human amenity of natural drainage ways.

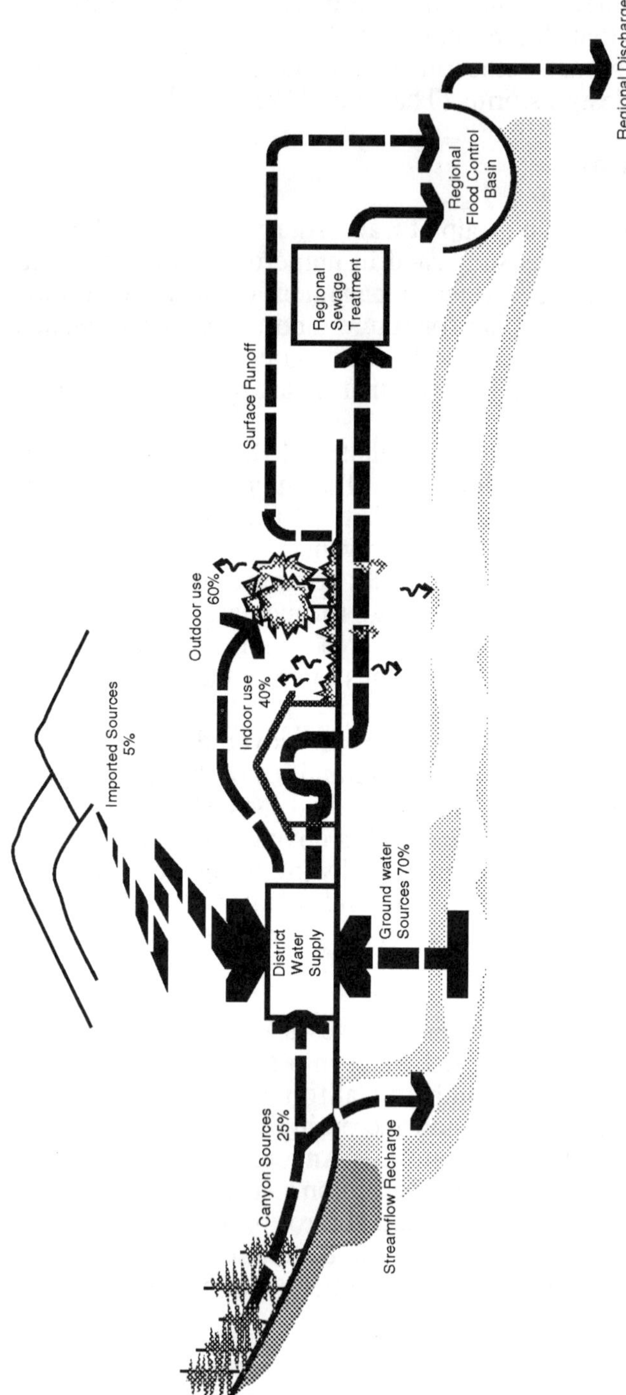

Figure 7.9 Hydrologic system planned for the Etiwanda-Day area, California (after Carlson et al., 1986).

In contrast, the Etiwanda-Day plan appealed to natural on-site hydrologic processes and to water conservation and reuse for sustaining plantings, supporting urban residents and replenishing ground water supplies. Figure 7.9 shows the conceptual hydrologic system. The canyon watersheds were to be preserved for continued water supply. Existing sand and gravel pits were to be used as recharge basins following mining. Proposed residences were clustered to leave permeable open spaces and drainage ways between groups of homes. On-site runoff was to be reused for irrigation during small, frequent storms. During large storms porous asphalt and open-celled pavers were to eliminate some runoff at its source. Dry wells were to be placed at building downspouts. Infiltration trenches were to intercept sheet runoff below roof lines and next to parking areas. Excess runoff was to be conveyed by natural drainage ways or channels lined with grass or loose rock to multiple-use detention areas. Check dams, percolation ponds and spreading grounds were to be placed along drainage ways of all sizes to reduce channel erosion and infiltrate water. Reclaimed wastewater was to be used on-site for irrigation of golf courses, agricultural lands, plant nurseries, and parks.

These approaches can minimize the hazardous effects of runoff, maintain the value of water as a natural resource, and take advantage of the ecological, visual and recreational qualities of drainage systems. Their applications can be adapted to a site's specific soils, geology, slopes and vegetation. While protecting lives and property, recreation can be provided, quality of life can be maintained, cost savings are possible, and the hydrologic cycle can be balanced.

REFERENCES

Aronson, D.A., and G.E. Seaburn, 1974, *Appraisal of the Operating Efficiency of Recharge Basins on Long Island, New York in 1969*, Water-Supply Paper 2001-D, Washington, D.C.: U.S. Geological Survey.

Berg, Vincent H., and Paul F. Clement, 1992, Maryland's Stormwater Park Brings Control to Runoff, *Land and Water* vol. 36, Nov.-Dec. 1992, pages 26-31.

Breeding, N.K., Jr., and J.W. Dawson, 1977, Pros and Cons of Storm Water Recharge Wells, *Water and Sewage Works* vol. 124, no. 2, pages 82-84.

Buekeboom, Th. J., 1982, A Dry-Well System for Excess Rainwater Discharge, pages 205-207 of 1982 *International Symposium on Urban Hydrology, Hydraulics and Sediment Control*, Lexington: University of Kentucky.

Carlson, William, Derrik Eichelberger, Rosa Laveaga and Alan Wong, 1986, *Design for the Etiwanda/Day Canyon Area*, Pomona: California State Polytechnic University Institute for Environmental Design.

Corbett, Michael N., 1981, *A Better Place to Live, New Designs for Tomorrow's Communities*, Emmaus: Rodale Press.

Ferguson, Bruce K., and Thomas N. Debo, 1990, *On-Site Stormwater Management: Applications for Landscape and Engineering*, second edition, New York: Van Nostrand Reinhold.

Hannon, Joseph B., 1980, *Underground Disposal of Storm Water Runoff, Design Guidelines Manual*, FHWA-TS-80-218, Washington, D.C.: U.S. Federal Highway Administration.

Jackura, Kenneth A., 1980, *Infiltration Drainage of Highway Surface Water*, FHWA/CA/TL-80/04, Washington, D.C.: Federal Highway Administration and Sacramento: California Department of Transportation.

Jensen, Ric, 1990, Storing Water Underground, *Texas Water Resources* vol. 16, no. 4, College Station: Texas Water Resources Institute.

Johnson Sustronk Weinstein and Associates, 1983, *South-East Transitway / Riverside Drive, Rideau River Park, Ottawa, Ontario, Environmental Impact of Filling and Grading, Phase I (Pilot Project)*, Ottawa: Regional Municipality of Ottawa-Carleton Transportation Department.

Jones, D. Earl, 1967, Urban Hydrology—A Redirection, *Civil Engineering* Aug. 1967, pages 58-62.

Juneja, Narendra, 1974, *Medford: Performance Requirements for the Maintenance of Social Values Represented by the Natural Environment of Medford Township, N.J.*, Philadelphia: University of Pennsylvania Department of Landscape Architecture and Regional Planning, Center for Ecological Research in Planning and Design.

Ku, Henry F.H., and Dale L. Simmons, 1986, *Effect of Urban Stormwater Runoff on Ground Water Beneath Recharge Basins on Long Island, New York*, Water-Resources Investigations Report 85-4088, Syosset, New York: U.S. Geological Survey.

Lindsey, Greg, Les Roberts and William Page, 1992, Maintenance in Stormwater BMPs in Four Maryland Counties: A Status Report, *Journal of Soil and Water Conservation* vol. 47, no. 5, pages 417-422.

Livingston, Eric H., and Ellen McCarron, (no date), *Stormwater Management, A Guide for Floridians*, Tallahassee: Florida Department of Environmental Regulation, Stormwater/Nonpoint Source Management.

Logan, Homer H., 1990, Flood-Control Dams Provide Water for Recharge, pages 499-503 of *Proceedings of Conserv '90*, Dublin, Ohio: National Water Well Association.

Maryland Water Resources Administration, 1984, *Standards and Specifications for Infiltration Practices*, Annapolis: Maryland Department of Natural Resources, Water Resources Administration, Stormwater Management Division.

McHarg, Ian L., and Jonathan Sutton, 1975, Ecological Plumbing for the Texas Coastal Plain, *Landscape Architecture* vol. 65, Jan. 1975, pages 78-79.

Muckel, Dean C., 1959, *Replenishment of Ground Water Supplies by Artificial Means*, Technical Bulletin 1193, Washington, D.C.: U.S. Department of Agriculture, Agricultural Research Service.

Newville, Jack, 1967, *New Engineering Concepts in Community Development*, Technical Bulletin 59, Washington, D.C.: Urban Land Institute.

Nightingale, Harry I., 1978, Lead, Zinc, and Copper in Soils of Urban Storm-Runoff Retention Basins, *Journal of the American Water Works Association* vol. 87, no. 8.

Nightingale, Harry I., 1987, Organic Pollutants in Soils of Retention and Recharge Basins Receiving Urban Runoff Water, *Soil Science* vol. 144, no. 5, pages 373-382.

Nightingale, Harry I., 1988, Artificial Recharge of Urban Storm-Water Runoff, pages 211-219 of *Artificial Recharge of Ground Water*, Symposium, IR Division, American Society of Civil Engineers, Anaheim, California, August 1988.

O'Hare, Margaret P., Deborah M. Fairchild, Paris A. Hajali and Larry W. Canter, 1986, *Artificial Recharge of Ground Water, Status and Potential in the Contiguous United States*, Chelsea: Lewis Publishers.

Palmer, Arthur E., 1981, *Toward Eden*, Winterville, N.C.: Creative Resource Systems.

Pensyl, L. Kenneth, and Paul F. Clement, 1987, *Results of the State of Maryland Infiltration Practices Survey*, Annapolis: Maryland Department of the Environment, Sediment and Stormwater Division.

Peterson, Frank L., and David R. Hargis, 1973, Subsurface Disposal of Storm Runoff, *Journal of the Water Pollution Control Federation* vol. 45, no. 8, pages 1663-1670.

Pettyjohn, Wayne A., 1981, *Introduction to Artificial Ground-Water Recharge*, Ada, Oklahoma: U.S. Environmental Protection Agency, Office of Research and Development, Robert S. Kerr Environmental Research Laboratory.

Seaburn, G.E., and D.A. Aronson, 1974, *Influence of Recharge Basins on the Hydrology of Nassau and Suffolk Counties, Long Island, N.Y.*, Water-Supply Paper 2031, Washington, D.C.: U.S. Geological Survey.

Sutton, Jonathan, 1974, Hydrological Balancing Act on a Texas New Town Site, *Landscape Architecture* vol. 65, Oct. 1974, pages 394-395.

Thayer, Robert L., Jr., 1977, Designing An Experimental Solar Community, The Village Homes: Phase 1, *Landscape Architecture* vol. 67, no. 3, pages 223-228.

Thayer, Robert L., Jr., and Tricia Westbrook, 1989, Open Drainage Systems for Residential Communities: Case Studies from California's Central Valley, pages 152-160 of *CELA 89: Water, Proceedings of the 1989 Annual Conference of the Council of Educators in Landscape Architecture*, Washington, D.C.: Landscape Architecture Foundation.

Weaver, Robert J., 1969, *Recharge Basins for Disposal of Highway Storm Drainage: Theory, Design Procedure, and Recommended Engineering Practices*, Research Report 69-2, Albany: New York State Department of Transportation, Engineering Research and Development Bureau.

Zwingle, Erla, 1993, Wellspring of the High Plains, *National Geographic* vol. 183, no. 3, pages 80-109.

Index